PROFIT FROM FACEBOOK ADS WITH $5 A DAY

HOW YOU AS A BUSINESS OWNER CAN BUILD A FOLLOWING OF LOYAL CUSTOMERS AND AVOID AD REJECTION, EVEN IF YOU ARE A BEGINNER

JEFF MURRAY

CONTENTS

INTRODUCTION

A third of the world's population are registered Facebook users. Imagine what that can do for your business. Facebook provides your business with a great opportunity to start and develop your brand with no risk and little to no cost. This is only achievable if you are informed of what has to be done to effectively advertise on Facebook. Want to learn more about Facebook advertising strategies? Do you want to learn how to create, run, and analyze advertising campaigns? How about using Facebook's resources to help you build your online brand? If this is the case, I believe this book will ultimately support you in doing so, especially if you're new to Facebook Marketing.

Most of us have Facebook, but very few people know how to use it effectively to advertise products, services, and even personas. With the help of this book, I will equip you with all the information to flourish in the realm of Facebook. You will learn the fundamentals, how and when to use Facebook Ads Manager, the steps involved in Facebook advertising, the various analytical tools available to measure your growth, how to use Facebook Pixel, metrics to track, and a variety of different tips for creating high-quality content. You don't have to be concerned if you don't have any previous expertise or experience in internet or social media marketing.

This book will serve as a guide to ensure you reach your Facebook marketing objectives and discover how to sell your items on Facebook. So, if you're keen to learn more, let's get started right now. Potential customers are now looking for companies like yours on Facebook. An unmistakably focused Facebook advertising method is the greatest technique to capitalize on this present audience. According to Facebook, your Business Page is "the cornerstone of your online persona." This book provides my most useful Facebook strategies for you to use.

The author, Jeff Murray, originates from Fairbanks, Alaska, and currently lives in Columbia, South Carolina. He started his entrepreneurial journey in 2014. His passions include fishing, fitness and, most of all, technology. Through years of research and experience in Facebook Marketing, Jeff has generated countless sales for his business, using the methods explained in this book. He knows the struggle is real, and he also wants to help other businesses by sharing his expertise and passion for Facebook Marketing. What he has learned took him years to achieve. Therefore he would like to share his knowledge about achieving sales and customers through Facebook with others.

1

TRADITIONAL MARKETING VERSUS FACEBOOK MARKETING

There are two types of marketing: digital and traditional. Traditional marketing has been around for quite a long time. The internet has altered its role in the world. There are many new ways to advertise online now, kudos to digital and online marketing. Social media marketing is just one of many ways to do it but because both traditional and social media marketing have the same aim, it can be a little hard to tell them apart. What's the difference between them, and which one should you use? It's an age-old question: How do you choose the right type of advertising? Things have changed a lot over time, and also the way people think about it.

WHAT IS DIGITAL MARKETING?

Digital marketing is using online platforms like websites and social media to spread the word about your business. If you use social media, you'll know all about the ads that show up in your stream. These are examples of digital marketing. When there are over 4 billion people on the internet, it's rare for them not to look up a product brand online before they buy it. It's also important to note that social media plays a big role in product research.

Channels for digital marketing include:

- Social media (Twitter, Facebook, Instagram, LinkedIn, Snapchat, etc.)
- Content marketing on a website
- Affiliate marketing
- Direct Email marketing
- Inbound marketing
- Pay per each click (PPC) advertising
- Search engine marketing (SEM)

WHAT DOES TRADITIONAL MARKETING LOOK LIKE?

Traditional marketing uses things like billboards, television, and printed media to grab people's attention.

Traditional Marketing Channels include:

- Outdoor advertising (billboards, bus and taxi ads, posters, etc.)
- Broadcast (TV, Radio, etc.)
- Print outs (Magazines, newspapers etc.)
- Direct mail (catalogs, letters, etc.)
- Telemarketing is when people try to get you to buy (Phone, text message)

- People can see through the window and see signs.

WHAT'S THE DIFFERENCE BETWEEN DIGITAL AND TRADITIONAL MARKETING?

The major difference between digital and traditional marketing is how people receive the message. While traditional marketing uses traditional media like printed magazines and newspapers, digital marketing uses digital media like social media or websites.

Even if traditional marketing is old school, it still applies to modern businesses. Traditional marketing fulfills a key role in people's lives because they do sometimes need to get away from the digital world. As important as they were two decades ago, interactive TV commercials and tactile Rolling Stone magazines are still important today because they help you remember what you saw. They will be at the top of your mind because you will connect emotionally.

In the same way, digital marketing is as essential as traditional marketing, if not even more important. This is how digital marketing works: It takes advantage of everything you do online every day so it can get to know you better.

If you search for ideas for your next weekend getaway on Google, you're likely to see an ad from a travel agency soon after.

By cleverly incorporating marketing messages into all the digital channels, digital marketing makes use of the fact that most people spend most of their day online.

How Should You Market Yourself or Your Business?

The best way to run a good marketing campaign is to find the right mix of traditional and digital marketing. When both are used, it is like Yin and Yang in a marketing plan. A good example is the Guinness approach. Their first ad (click here to view) that appeared in the Daily Express in February 1929, promoting Guinness Brewery, is still available to view on their website. For the unique and powerful way they film their TV commercials, they're known as "the best." The Guinness 'Surfer' ad campaign from 1999 is still one of the best TV ads of all time. It doesn't matter that Guinness is famous. They still need to keep up with the times and use digital marketing to make sure they don't miss out on some great marketing opportunities. In order to reach more people and a younger following, Guinness has recently expanded their digital marketing reach by making videos for Facebook and Instagram. Instead of just re-editing their TV commercials, they filmed their campaigns as social first videos by putting together the shots with both Facebook and Instagram in mind when they made it. They made the videos for the people they had in mind. The result was a visually stunning and powerful social media campaign about cowboys who take care of their horses.

ADVANTAGES AND DISADVANTAGES OF TRADITIONAL AND DIGITAL MARKETING

There are a lot of differences between digital and traditional marketing, so let's get down to the nitty-gritty and figure out what they are and how they work for and against each other.

The Good and The Bad of Traditional Marketing

If you have the money to put your ads in magazines and on TV at the right time, you could spend your money wisely.

+ Advantages:

- It's powerful and easy to understand.
- It is normal for people to see a bold billboard or a strong TV commercial in their day-to-day lives. They're easy to read and can be fun.
- Marketing materials that are printed are more durable.
- If you put an ad in the New York Times, it will stay there until the magazine gets thrown away.

- Traditional marketing is great if the customer is a big fan of collecting things like promotional gifts.
- Seeing something in person rather than on your phone is more likely to help you remember it.
- Super Bowl ads or a beautiful and impressive window display are more likely to stay in your head than a Facebook ad that you'll probably scroll past in a few seconds, though.

— **Disadvantages:**

- Difficult and expensive to measure campaign outcomes.
- Expensive for upcoming businesses.
- No direct interaction with the customer.
- No 24/7 direct engagement where you know what your audience thinks.
- Even if traditional marketing is powerful, we can't forget that we're living in the digital age right now.

According to ClickZ, Internet users now make up 57% of the world's population, and daily users viewing Facebook stories have increased to 500 million (in 2019). People spend an average of 6 hours and 42 minutes on the internet each day, but this can vary from person to person. Some spend less, and others more. As of 2021, 73% of all ecommerce sales were done on a mobile phone. A lot of time and opportunity to get clever with digital marketing.

Digital Marketing and Its Ups and Downs

✛ Advantages:

- More ways to get involved and engage with clients.
- A far greater reach, you can do business globally if you prefer.
- What people think about your brand's marketing initiatives is available right in front of you.
- If your content post was shared, liked, and received positive comments, you know you're doing something right.
- Measure your campaigns with ease: It is simple to track your campaigns.
- The precise details of digital marketing tracking are much more in-depth than traditional marketing.
- This makes your lessons very clear for your next round of marketing.
- Makes it possible to be clever about where you target your ads.

- If you have the tools to target a 29-year-old female writer who likes Lizzo and Guinness, then surely you can also come up with specific content that is just right for her. Get creative with your reach.

— **Disadvantages:**

- Many people don't like digital ads that pop up when they are online. Think about when you're on Facebook and a sponsored ad comes up for something you looked up the night before about an embarrassing health problem. It's almost certain to make you hate the brand that did the clever targeting.
- Some digital marketing efforts are only there for a short period, like Google ads, banners, promotional emails, and social media ads. They aren't tangible, so it's easy to forget about them.
- Users keep scrolling or clicking to the next page, and your ad will be gone from their screen.
- Digital channels and media always change. To get as much out of your online marketing, you need to learn a lot.
- People who specialize in one thing usually specialize in another thing. For example, search

engine marketing and social media both need a pro to get the most for your money. This can be expensive.

The most important thing is to figure out what your needs are, how much money you have, and who your target audience is.

A study from 2019 (GfK) says that 56% of people say that TV helps them learn about new products. In addition, the study found that TV ads led 86% of adults and 91% of millennials to look for a brand or product online.

Social media does not fall far behind. A modern consumer uses a lot of different ways to investigate a product or service, and checking the brand's social media accounts is a must. Here, if they like the brand's social media content and how it interacts with its fans, they will be more likely to buy from this business.

If you want to start with social media marketing, start with a basic strategy. This book will provide you with good ideas for your Facebook advertising strategy. Find what works best for you.

Digital marketing isn't just for millennials and Gen Z'ers. My grandmother spends hours on Facebook, watching videos, and shopping on the internet. Thus, you should know your audience. If you feel you're connected to a brand, you're more likely to choose it over its competitors. There is a lot of research that shows that in getting people to talk to you, social media marketing is better than other marketing.

Both digital and traditional marketing can deliver results for you, if you know what your target market needs and how to reach them. A combination of both may create the ultimate customer experience.

The media and channels used by both fields are good ways to get people to buy products and services. If you want to see results quickly, it's a good idea to mix social media strategies with some traditional marketing tactics to get your business more exposure.

2

GETTING STARTED ON FACEBOOK

WHY USE FACEBOOK?

With 2.82 billion people who use Facebook every day, it's important to know how to use social media to market your business (Beveridge, 2019). We conclude the following benefits from what we have discussed regarding digital marketing:

- It's easier to target your specific audience.
- The budget is more scalable.
- Brand awareness.
- Digital marketing is not just limited only to online businesses.
- Directly communicates with groups or individual consumers.

For most marketers, Facebook's organic reach has dropped over the last few years. However, that doesn't mean that you should give up on this social network right away. Not at all! Marketers say that Facebook is still the best social media marketing tool out there, according to a recent poll (Hootsuite, 2021). According to 95.8% of people who took part in a report called The Future of Social Marketing, Facebook was one of the top three social networks for getting a good return on investment (ROI). Because of this, 63.5% of respondents said that Twitter and 40.1% said that Instagram were two of the top three ways to get a good ROI.

Important Reasons Why Facebook Is the Best Way to Market on Social Media

Facebook is the most popular social media site out there. In social media, Facebook is the 800-pound gorilla. It has over 2 billion users, and over 900 million of those users visit the site every day. As a small business, you can't afford not to have a presence there. Facebook users come from all walks of life.

- Whether you want to reach teenagers or old people, you can find them on Facebook. According to Pew's website, about 72% of people who use the internet in the United States use Facebook. Most online adults ages 18 to 29 use it, 79% of those ages 30 to 49 use it, and 64% of those ages 50 to 64 use it. Even people who use the internet are 65 and older and use Facebook, nearly half of them (48%) do.
- Many people spend a lot of time on Facebook: 40 minutes a day, to be exact. Business Insider (2015) says that around the world, people spend 20% of their time online on Facebook. Your business will spend more time on Facebook when you spend more time there.
- Facebook advertising is very specific. You can target your audience based on their location, age, gender, interests, and many other things. Target existing customers with the Custom Audiences feature in a safe and private way.
- Facebook lets you reach out to people who have bought things from your website through the site. When you use Facebook's re-targeting option, you can only show Facebook ads to

people who have already visited your business website.

- To spread the word on Facebook, you can send a message to all of your friends' friends. It's a lot more effective when your customers see your Facebook posts or ads because their Facebook friends see them, too.
- Facebook helps you keep track of your progress.

Are you still not convinced that Facebook is a good way to market your business? Measure the ROI. If you want to see how your Facebook marketing is working, a free tool called Facebook Insights can help. It lets you see how your ads and organic posts are working.

You can get a lot of information about how people are interested in your business on Facebook, like how many likes you got this week and how that compares to how many likes you got in the past, how many comments and shares you got on posts, and which ads led people to your website or other actions.

To be able to market your business effectively, you need to know how Facebook is different from other media and how it can help you. Just like you wouldn't run a radio ad on TV, you shouldn't run an ad on Facebook the same way you would on TV.

Don't use Facebook to try to "hard sell" people. People think of Facebook as a fun place where they can talk to their friends, look at photos and videos, and just chill out. You need to join in on conversations and become part of a group, rather than being a business 'outsider' who tries to sell aggressively to people in the group.

Hard-sell tactics, like using advertising slogans, posting about a product or service over and over, or giving out lists of products and prices without any conversation, will make other people 'unfollow' you. They might even write bad things about your business.

What Makes Facebook Different from Other Online Platforms

- Facebook's ad platform has budget management tools that are easy to use, near real-time results monitoring, is available to any business, no matter how big or small, and can drive measurable results, like revenue. Finally, there is also an indirect benefit to keeping your Facebook presence. This is because it helps your organic search rankings if you keep it up.
- Facebook advertising can help you get more social engagement (shares, likes, and comments), which can help your SEO rankings in the long run.

- Before COVID-19 started, people were already spending an average of 58.5 minutes a day on Facebook. During the pandemic, Facebook and other social media use has gone up even more. Now is the best time to use this channel to help you advertise. Facebook is the most popular social media platform with more than 2 billion visitors daily.
- Facebook advertising is affordable. Cost about $6 per 1000 impressions. As a bonus, you can set and control your own budget, so you don't have to pay for extra marketing costs.
- Talk to existing and potential customers—by posting and receiving messages, and posting images and videos to start engagement.
- Targeted advertising to a specific audience: age, location, interests, etc...
- Drive traffic directly to your Site—If the ad provides enough relevancy and value, the user will leave Facebook and proceed to your site.

In 2021, Facebook changed its name to Meta. Meta is Facebook's parent company, and it runs Instagram, WhatsApp, and Messenger. For marketers, this provides even more opportunities. Meta has made $117.9 billion in total. Not bad at all for a company that started in a Harvard dorm room. A lot of this money came from Meta's apps. Yes, this includes Facebook, and this number has only gone up each quarter as more people see the value in scrolling through Facebook, Instagram, and WhatsApp. If you're on Facebook, you're more than likely to use the other apps in Meta's family of apps, too. Many people who use Facebook also use YouTube. It turns out that 71.2% of people who have Facebook also use WhatsApp. Instagram is used by 78.1% of Facebook users, and a lot of other people also use it. People who use Facebook are less likely to use TikTok and Snapchat, which are more popular with younger people.

In the United States, 70% of adults use Facebook. It's the most popular social media site for men and women 35 to 44 years old. Men make up more than half of Facebook's ad audience, and women make up the rest according to Facebook's advertising audience report (Hootsuite Digital Trends Report, 2021).

According to research by Pew, no other major platform comes close to this level of use. Only YouTube, which is used by 80% of Americans. Most Americans say that they go to Facebook several times a day. A big part of Facebook's success is because more people visit the site, which means they have a better chance of seeing ads.

WHAT IS A FACEBOOK BUSINESS PAGE AND WHY CREATE ONE?

To set up a Facebook Page for your business, you don't have to spend any money. Anyone looking to build an audience on Facebook can create an account. This is great for marketing through posting and sharing content. It allows an unlimited amount of followers versus the Facebook friends limit. It is also a requirement to post ads which will be explained later.

"A place for any public figure or group to have a conversation with Facebook users... a public profile that helps users connect to what they care about." This is how Facebook describes its Pages used by musicians, artists, sports teams, companies, media, nonprofit groups, and more.

Each page asks the user to set up the menus, add basic information, and check to see if fans can post on the site, upload photos and videos, and do other things to keep things safe. Then, they even send you metrics (information, fans, and posts) called 'Insights' that you can't see on your own profile, like this: Create a Facebook page for you. The first thing you need to do is set up and categorize your Facebook Page. Starting at the top of your Facebook home page, click on the Pages link on the left. This will open the Pages menu. Pages you like or follow show up here. You can also see who is in charge of these pages. This is how the dashboard looks from this point of view. Every day, it tells you how many notifications, how many likes, and how many active fans you have each day. This is also where you will start by setting up your page.

Start Setting Up Your Page

After signing up as a Facebook user, go to www.facebook.com/pages/create. Select the type of page you want to create: business/brand or community/public figure. People won't see your page until after you've named it and set up all the tools.

Complete your Page information: page name, category and description. Click "Create Page" when you are done. Decide on a Page name that you like! You can't change the name after you have created it. You'll have to delete it and start over again.

You'll have to give specific information about yourself. Local businesses and individuals want to know more about the person they are buying from and establish a trusted online relationship.

Make sure that your page is in the correct category. Add your profile and cover images—use photos that best represent your brand (i.e., logo, signature product, or yourself if it is a personal entity).

Your profile image should be 170 x 170 pixels. It will be cropped to a circle, so don't put any critical details in the corners. Click "add profile picture once complete."

Choose a Cover Image that illustrates what your brand represents. The recommended image size is 1640 x 856 pixels. Then click 'Add Cover Photo.'

▷ **Creating a Username**

Also called a vanity URL, and you can use this as the best method for discovering you. Despite being allowed 50 characters for a username, I recommend using a username that is easy to remember and recognizable. Once you have received the green check from Facebook, you are all set. Go ahead and create the username.

▷ **Add Business Details**

Follow the steps provided. If you have a website, enter it when prompted. Enter business hours. Add a CTA or call to action. A built-in call-to-action (CTA) button on Facebook makes it simple to provide the customer with exactly what they're searching for and gives them the opportunity to interact with your business in real-time. Customers will be encouraged to discover more about your products or services if you use the proper CTA button. To add a CTA, click the blue Add button, then choose the required action. You can opt to unpublish your Facebook Business Page if you wish to take it down while you decide about the details. Click Settings, then General, from the Manage Page menu. Change the status of the page to unpublished by clicking "Page Visibility."

▷ **Create Your First Post**

Ensure the content is valuable. Either create your own or share from others in your industry. You will learn more in the next chapter about how to post valuable content.

▷ **Customize the Settings**

Also, Facebook lets you change other preferences that are near your personal pages. Take your time and change your mind as you go through each page. The first view of your new Page lets you do six things:

- Add a picture to your post.
- Invite your Facebook friends to come over and have a good time with you.
- You can tell your fans by adding their contact information.
- Share information about your life.
- Put a 'Like' box on your website to show people how much they like the page.
- Make sure your phone is set up so that you can update the page from anywhere.

▷ Add Your Logo

It is because your Facebook page extends to your business that you must use your company's logo (or headshot if you're just you) as the picture that shows up when people first look at your page. This logo can be 200 pixels wide and 600 pixels high, so it can be used for a lot of things. Take your time when choosing the thumbnail, because this version of the thumbnail will show up next to all of your posts.

▷ Upload Pictures and Videos to Your Page

No, I don't have any pictures of my office, my workers, or anything else that is connected to my business. Well, it's time to up your game then! Make a few photo albums and share them with everyone. Keep in mind that you're not just trying to build a brand on Facebook, but also to make your company seem more human to people. One of the easiest ways to show this is to show the people and places that make the company work every day. If your office needs to be changed, it might also be a good reason to change things up a little.

In fact, if you have pictures of your products or a screen grab of your software, then you should make an archive for these things. An overview of each picture or snapshot can be given. This can be done in a few words.

Uploading video content on a digital platform takes more time than putting a picture on one. You can share another kind of media with your clients, buyers, fans, and even other people who are interested in your business. Because these videos aren't on your company website right now, you might not be able to show them later. On your Facebook page, you can put together pictures that you have taken from all over the world. People can buy products over Facebook by seeing the picture and price. You may even want to create a shop on your Facebook page. Many sites don't show the Videos option right away in their browsers. You might need to go to the Edit Video Page or the Applications Page to make changes to your video. To change the settings for your video, choose (add) next to Open in the video settings.

▷ Add Useful Content

People who use your products or services have given you feedback. You can also show off the content of your projects. Ads or meetings behind the scenes. Which would you rather see? Use the publisher tool to put them on your site.

Because Facebook is all about the human experience, using it only as a commercial tool will only get you so far. Post news and updates in a simple, casual, and enjoyable manner. There will be no marketing gimmicks or press releases!

Just that little self-promotion won't hurt when it comes to Facebook achievements. Share the news when you get 1,000 followers or 100 or 200 likes on a post. This gives them the impression that they are a part of a developing community.

Always include a link to your company's website and other social media pages, for example Twitter, LinkedIn, and Instagram. This improves the search engine rankings of your page, improving your potential presence with prospective clients.

Content should be mixed between posts, videos, photographs, competitions, polls, and so on. Keep things interesting and surprising!

If you run out of ideas, browse through your archives for pieces or comments from fans that may be recycled or given a fresh twist. Posts tying your company to current events are in most cases well-received by Facebook users.

By remaining focused on a crucial principle—Facebook is less about marketing and sales and more about dialogues and relationships—you'll be able to generate a plethora of content ideas that will help you establish a growing and passionate community of followers.

If your business is on Twitter, you will find apps that let you build your own Twitter presence. It has its own Twitter feed tab on your site. Flickr and YouTube have apps, too. You can also use them to take pictures and make videos. You may also consider sites like Poll-Daddy to add polls to your website, blog, or Twitter, calendar publishing software, and a million other tools to help you with your online presence.

How to Connect Your Page with Other People

You need to start building your audience. A powerful way to communicate with customers and people who might be interested in your business is to let them see your product or services and give them the option by adding an action button to your posts to send you a direct message via Facebook Messenger or WhatsApp.

Facebook also lets people 'tag' photos so they can tell if a friend from Facebook is in them. If you run a business, you can use this feature to spread the word about your company

For example, a tour operator could put a picture on their Facebook page of a group of people going white-water rafting. They could then ask each person's permission to tag them in the picture. It will appear as an update on the involved persons Facebook profiles, where their friends will see it. This makes the picture and your business more interesting. If you do decide to use tagging, be very careful about how you do it because it can be a personal privacy issue, and some Facebook users are afraid of being named in photos.

If you share information about your business that other people will find useful or interesting, you'll have a much better chance of making money. This increases your credibility and helps your business because it helps you build long-term relationships with other people who use your service.

▷ Begin by Inviting Your Existing Facebook Friends to Like Your Page

To start, scroll down to the bottom of the "Set Your Page Up" box and expand the item titled "Introduce Your Page." To bring up a list of your personal Facebook friends, click the blue "Invite Friends" button. Choose the friends you wish to invite, then click the "Send Invites" button. Use your existing platforms, such as your website and Twitter, to advertise your new page. Include "follow us" logos in your advertising materials and email signature. If you feel comfortable doing so, you can invite your customers to leave a review on Facebook.

▷ Connect to Other Pages

Because Facebook is a social network, using your Page to establish a community for your business is a fantastic idea. Connecting with other Pages that are related to your business is one approach to establishing a community (but not competitors). For example, if you own a store in a busy shopping district or mall, you may network with other stores in the same location. Consider it an internet equivalent of your local chamber of commerce or business improvement organization. If you run a virtual business, you might network with other companies in your field to give additional value to your followers without directly competing with your products. To follow other companies on Facebook, go to their page and click the more icon (three dots) under the cover photo. As Your Page, choose Like.

▷ As Your Page, Join Groups

Facebook Groups provide an organic way to reach a large number of individuals who are interested in a given issue without having to pay for advertisements. Joining and posting to a relevant Group as your Facebook Page encourages anyone who is interested in your post to visit your business page rather than your personal profile.

Why Don't People Know About My Page?

After recruiting your first 25 page likes, the thing you should do next is definitely make a URL that is easy to remember that leads to your Facebook page. You will make a unique URL from scratch. Facebook has what's called a 'vanity URL' which means that the URL is only for you. For example: http:/facebook.com/YourBrand.

▷ How to Create a Unique Facebook Business Page URL

When you create your Facebook business page, you may realize that the URL is a little, well, unattractive. It's also lengthy, making it tough for individuals to remember. The first URL that Facebook assigns to a business page will have a series of digits after the www.facebook.com/ that won't identify your brand. A lengthy URL makes it tough to describe how to locate your Facebook page to others, and it will seem unprofessional on your business card and promotional materials. Fortunately, Facebook has a solution for those of us looking for a shorter, simpler version of the link: a custom Facebook business page URL (also known as a 'vanity' URL or, as Facebook refers to it, a 'username').

We'll show you how to create a custom URL for your Facebook business page! The days of having a Facebook URL the length of your arm are over. To create your own Facebook business page URL, follow these steps:

1. Go over to your business page and click the "Create Page @Username" button on the left side of the page.
2. Create a username and a custom URL.
3. Click "Create Username" if the username isn't already taken.

Keep in mind Facebook's rules for generating a user-name while selecting a custom URL.

Having a personalized Facebook business page URL brings you one step closer to fully using your social profiles. You and your followers/potential buyers will find it simpler to remember your personalized URL now that you have one! Because Facebook has billions of users, the sooner you can establish a new, unique username for your page, the better—usernames are rapidly grabbed, particularly if the name is popular. Always keep in mind that your unique URL should be distinctive and relevant to your company.

We hope this information will assist your company in getting off to a good start on Facebook!

Building a Sales Funnel for Facebook

A Facebook Ads funnel is a series of ad campaigns meant to guide people from total strangers to paying clients. The Facebook Ads funnel is made up of numerous campaigns with multiple ads that feature various value propositions and are delivered to the correct audience at the right time.

A mix of organic content and Facebook ads will allow you to reach the largest potential audience. To strike an optimum balance, you'll need a sales funnel created explicitly for Facebook. Before we get into how to create one, let's first explain what a Facebook sales funnel is.

A Facebook sales funnel is a route that leads Facebook users to a specific activity, in this example, making a Facebook purchase. Unlike a typical sales funnel, it is particularly intended to convert Facebook users. The funnel is separated into three sections, as shown below. Let us take a deeper look at each component.

The top of the funnel (ToFu) is where you try to increase brand recognition among your target audience. This may include developing adverts that quickly explain your items and promote them to look-alike audiences, holding a referral contest, and providing a range of material—such as blog posts, videos, infographics, and ebooks—to appeal to diverse portions of your audience.

The funnel's core is the next level of the Facebook sales funnel (MoFu). Throughout this stage, you should concentrate on generating leads and nurturing these leads.

You could answer questions in the comments section of your ads and posts, push out content that details your products in greater depth and has performed well in the past, and run retargeting campaigns to show customers how you're different and why they should buy from you. A video ad campaign targeting viewers who have previously clicked on a blog post or another sort of content from your Facebook page may be an example of such a campaign.

The funnel's bottom is the final level of the Facebook sales funnel (BoFu). Your major focus at this time will be ad conversion, followed by retention. As a result, the Facebook advertisements employed at this stage will be those that persuade your leads to purchase from you and then convert them into loyal customers.

As the last incentive for customers to make a purchase or take some action, these adverts may feature special deals or discounts, such as a free product or service coupon or trial (like opt-in to your newsletter). They should also be highly targeted in order to advertise items that users have already seen or complementary products based on users' buying history. Now that you have a good understanding of what a Facebook sales funnel is, let's talk about how to create a successful one for your company.

Steps for Building a Successful Sales Funnel

Facebook has a variety of ad options that you may utilize at various stages of your Facebook sales funnel. However, all of these alternatives might make the procedure complex for you too.

Let's go through the step-by-step procedure for constructing a Facebook sales funnel for your brand.

▷ Identify Your Hottest Target Market

Before you can begin utilizing Facebook to reach potential customers and develop your business, you must first determine who your brand's target audience is. Why, you may ask?

Because the material you provide must be tailored to their tastes. Targeted content has a higher chance of generating attention and converting customers.

Your target audience should be chosen based on a variety of demographic, geographic, and psychographic characteristics. It's also a good idea to develop buyer profiles for your target demographic and examine your sales process. As a consequence, you'll be able to better personalize your material for each character.

▷ Publish High-Quality Content

The foundation of every sales and marketing plan is content, and the same is true for a Facebook sales funnel.

You'll need to collaborate with your marketing team to develop segmented content that appeals to each buyer profile. A blog post, video, photograph, or even a slideshow might be used. What is important is that it be of excellent quality and relevant to the interests of your customer persona.

You should be extra cautious while producing material for blog articles since it must be devoid of plagiarism. To create original material, it is best to utilize a plagiarism detector such as Grammarly.

The goal is to capture your target audience's attention and encourage them to interact with your brand. The importance of good, tailored content in this effort cannot be overstated.

So, what should you do now? Make sure to vary your material to keep your audience interested.

Encourage people to participate with your material on Facebook, as well as share links to your blog entries, so you can drive them to your website. Remember that bringing them to your website may help you move them farther down your funnel.

You will, of course, need to determine which material produces the greatest outcomes for you. This may be accomplished by tracking its reach and interaction using Facebook's statistics.

Take a look at how HubSpot engages its Facebook fans using videos and graphics, for example. HubSpot's Facebook page serves as a sales funnel.

▷ **Use Facebook Ads to Increase Your Reach**

Once you've begun to see some interaction with your content, you should begin using Facebook advertisements to increase your reach and engagement even more. Because video is predicted to account for 82 percent of mobile traffic in 2021, video advertising will be especially successful.

It is critical to make good use of your advertising money. As a result, it's critical to start by promoting just your best-performing content.

But why is that? This is the most interacted-with content, thus it's most likely to do well as an ad as well. It is important to note that you should not advertise this information to your whole audience. Instead, concentrate on individuals who have previously shown interest in your company.

This is your warm audience, who are more likely to re-engage with your material since they are already interested in your company. Ideally, they will be your Facebook followers and those who have already visited your website.

You may add the Facebook Pixel on your website to monitor the traffic you've got from Facebook. It may display you all of the activities that users perform on your website after visiting it through Facebook.

Now that you've determined your target demographic for Facebook advertisements, you may push the content to a bigger audience. To handle these ads, you may utilize the Facebook Ads Manager or one of the various Facebook marketing tools.

You should analyze the statistics of the posts on a regular basis to see which sort of material is doing the best.

▷ Utilize Remarketing

One of the most effective methods to warm up a chilly audience is via content remarketing. You may be wondering how.

It allows you to introduce them to your brand many times. This may help to raise brand awareness. You can't expect people to interact with your material the first time they see it.

When people are constantly exposed to your brand, they are more inclined to interact with it, resulting in a warm audience for you. You should remarket to individuals who have interacted with your content in the same way that you market to those who have previously visited your website.

You may implement Facebook remarketing for your brand in a variety of ways, including:

- visits to websites
- interactions on events
- participation in your videos
- visits to websites
- combinations that are unique

Increased contact with your brand may persuade people to become leads or conversions.

▷ Target Look-alike Audiences

Look-alike audiences are one of the most useful tools provided by Facebook. It's a feature that allows you to target a similar audience to your present one.

This may take the uncertainty out of the equation and allow you to focus on your greatest audience with little effort.

When you give a source for the look-alike audience, Facebook will produce it automatically. A custom audience, a Facebook Pixel, or even the audience of a page might be the source.

You may also specify a place for the look-alike audience and a percentage of the people to target in that area ranging from one to ten percent.

To establish a look-alike audience for your business, go to the audience area of Facebook Ads Manager. Create a look-alike audience for your Facebook sales funnel.

When developing a look-alike audience for your business, keep the audience size minimal. The odds of the look-alike crowd matching your core target are higher with a smaller group.

▷ Be Proactive, Reactive

Another crucial consideration when creating your Facebook sales funnel is communicating with your audience in the medium they choose, which is rapidly becoming messaging applications. In fact, Facebook says that during the peak of the COVID-19 epidemic in 2020, total daily talks between individuals and companies on Messenger (as well as Instagram) increased by more than 40%. Remember that you must be available to your audience when they need it.

Respond to their comments and messages as soon as possible and fix all of their difficulties to gain their confidence and trustworthiness. You'll never be able to persuade them to purchase from you until you first win their confidence.

Also, if others see that you aren't handling your customers' concerns, they will be less likely to purchase from you, and vice versa. The objective is to warm up your audience by nurturing it.

You may use Facebook Saved Replies and Instant Replies to save time when addressing commonly requested questions. These Facebook functions are just one illustration of how artificial intelligence may aid in sales process optimization.

▷ **Offer Purchase Incentives**

Once you have a warm audience, all you have to do is prod them to buy from you. This may be accomplished by posting links to your product page.

Consider including discounts and offers in the posts to sweeten the deal. Making it a limited-time offer certainly helps to leverage on their fear of missing out (FOMO). This sense of urgency, paired with the discount, may aid in the generation of sales.

Furthermore, you should promote these articles to all people who have already interacted with your company. This increased reach within your warm audience might assist you in generating even more conversions.

For example, Inkkas has a Facebook ad with a 25% discount offer. The post directs you to their product page where you can finish your purchase.

▷ **Mobile-Optimize Every Stage**

According to Facebook's The Evolving Customer Experience study (2020), people are increasingly relying on mobile devices for both product research and purchase. Sixty-six percent of respondents said their mobile device is becoming their most significant purchasing tool, with 45 percent saying they purchased more on their smartphone during worldwide lockdowns.

That's why it's critical to mobile-optimize every level of your Facebook sales funnels. Facebook provides a few suggestions for mobile-optimized assets, including:

- converting static photos into slideshow advertisements
- reducing advert length to 10-15 seconds
- developing 9:16 adverts to suit mobile device vertical screen space
- overlaying text on videos so that people may watch with or without sound
- beginning and concluding with your company's or brand's message

▷ Concentrate on Customer Retention

The last phase in the Facebook sales funnel is to keep your clients committed to your business. It is critical to deliver excellent customer service to them by following customer service best practices. This ensures that they will continue to be happy with the services supplied to them. You should strive to solve all of their problems as quickly as possible.

It is also vital to motivate them to make further sales for you. This may be accomplished by:

- purchases in the future
- upselling
- programs for referring others

Upselling may be accomplished by conducting ad campaigns based on your customers' purchasing history. When consumers encounter comparable items or services, they may be tempted to purchase them. Similarly, for repeat transactions, you may retarget your consumers with fresh offers in order to entice them to make another purchase from you.

Ultimately, by updating people about your referral program and its benefits on Facebook, you may urge them to spread the news about your business. You may do this with both organic and paid posts. To build your audience and your following faster, the chapter that deals with ads will explain in detail.

Have a Clearly Defined Strategy

A Facebook page could help your business in many ways. It's true that some of these benefits are the same as having your own website, but many are unique to having an account on Facebook. As a whole, the benefits below can help your business make more money and sell more products.

It's critical to understand what you want to achieve using Facebook and how to act on it. Coffee shops might want to increase sales from Facebook by 10% in the next six months. This is an example of a goal. Their plan could include posting a picture of a customer who is "The Coffee King or Queen of the Day" every day and encouraging people to post their own coffee-themed photos. This way, the company can keep track of the sales.

For your Facebook marketing, you need to set a goal and a strategy. This will give you a direction and a way to measure your progress.

Business Events

You or your staff should set up the Events tab if your company hosts events, has webinars, or does other similar activities. Facebook groups should be set up for meetups, online events, and more. Make sure your Facebook page is listed as the organizer of the event. Facebook users can sign up for the event and share it in their news feeds. Streams are used to help spread the event's reach. Use an online registration service to sign up for an event, such as Eventbrite (eventbrite.com), then you can start setting up an event right away directly from Eventbrite. As such, you won't have to waste time posting the Eventbrite event on Facebook too.

Unless you're Justin Timberlake, you will have to work a bit harder to get the attention that your page deserves. Set your page visibility to 'Public.'

Provide Help to Customers

When you pay attention to what the public thinks about a business or product, a marketing campaign, or your industry you can learn a lot. Customers are able to ask you questions on your Facebook wall. Your staff can answer them there, too. This is often more efficient than having staff answer the phone, and it lets other customers read common questions and answers without having to talk to you one-on-one.

Brand awareness and good word of mouth can help you get more customers. Encourage customers who already like your business to click the 'Like' button on your Facebook page. This will raise your business's profile on Facebook and help you get more customers. Customers can write good reviews about your products or services where all of their friends can see it. Facebook can help you get more people to your website if you add a link to your site. Many businesses say that the most important thing about Facebook is that it directs new traffic to their site and the chance to buy goods and services increases.

The ads will be aimed at the people who are interested in them. Facebook can look at all of the information that millions of people put into their profiles and see how it works.

This information can be used by the owner of a business page. You can pay to use this information to send targeted advertising to a group of people.

People over a certain age and in a certain city might be able to use Facebook to figure out how many men in that city say they like fishing. Ads for new fishing lures could then be made and paid for only by people who saw them.

As soon as a person checks into a neighborhood, a street, or a business, they get a list of nearby businesses that are offering deals. This is called "Facebook Places" (e.g. discounts, freebies, loyalty rewards). Facebook Places lets people use their mobile phones to "check in" to a place so their friends can see where they are on Facebook. Facebook Places also tells you which popular places are near you when you check-in. Only pages that have been set up to be a Company and Organization or a Local Business can add a location to them.

Below follows important reasons why you should add Facebook to your marketing mix.

Getting Your Brand Noticed

Advertisers can't deny that Facebook has the most users. Statista (2020) published that Facebook had more than 2.6 billion users around the world, which is a lot. That's more than on any other social media site, and maybe even more than Google's search users. It's also more than any other social media site.

The people who use Facebook aren't just old people; they come from a wide range of ages and backgrounds. You should be able to find your ideal customer on Facebook, no matter what kind of business you run.

Facebook has a lot of younger users, with 62% of its users between the ages of 18 and 34. Facebook also has a lot of older users, with 38% of its users between 35 and 65+.

The fastest-growing group of Facebook users are people in their 30s and 40s. Baby Boomers (people born 1946-1964) have been on Facebook more and more since 2015, according to the Pew Research Center. People born before 1945 have also been on Facebook more and more.

Aligning with B2B and B2C Companies

Businesses that sell to both individuals and businesses should work together. Prepare to be surprised by how businesses that sell to other businesses can also run successful Facebook ads. Business leaders spend 74% more time on Facebook than the rest of us. The B2B market is very competitive, which means that B2B marketers need to be very aggressive when they use Facebook.

But with the right targeting, ad format, messaging, and off-Facebook user experience on your site, there's a good chance you can make money. Facebook remarketing is the last thing B2B marketers should think about. Everyone who is a B2B target doesn't stop being one when they leave the office or are online in snackable moments between work meetings. We often forget that. Remarketing to them on Facebook is a good way to stay in their minds.

A good B2B strategy is to build look-alike audiences based on an email list, website visitors, or customer base. Advertisers have seen it work well on Facebook, even though this isn't unique to Facebook. Other platforms now have this, too. People at Hawke Media used Facebook advertising to get a four times return on their ad spend with this first step.

This is a great time to hire vetted freelancers to help you grow your business. Fiverr Business gives your team the tools to work together and delegate with the world's best freelancers for any project.

Advertising: Full-Funnel Targeting with Many Different Types of Engagement

People who use Facebook may be the only digital platform that can help them at any point in their relationship with it. Facebook's ad formats, targeting options, and measurement tools work well with any marketing plan. It is useful whether the user is in the upper stages of the funnel, browsing and just starting their research, or ready to buy.

If you want to get people interested in your business without being too pushy, Facebook's sponsored stories, video ads, and carousel ads are great ways to do that without being too direct.

If the users aren't even in the mood to think about your message, they won't even look at it and will just move on.

As long as your customers aren't, this is your chance to shock and delight them with something that is very visual, unique, and shows your service or product in a way that is useful.

The goal is making the benefits attractive. The goal is to show something that will make the user excited and make them want to keep looking when the time is right for them.

To reach people in the middle of a funnel, Facebook has different ways to help you out. There are a lot of ways that you can use these tools to not only interact with people on Facebook, but also get them to do things off of Facebook that make them interact with content on your site.

If you want people to actively think about your business, then Facebook can be a great source of highly qualified traffic (or, if your business relies on app usage, to drive app installs).

Transactional marketers will benefit from conversion campaigns. Make sure to use the custom buttons that Facebook has to make your ads even better.

Facebook has a variety of buttons that you can use in your ads with a specific call to action, depending on your industry and your goals. These buttons can help you tailor your ad to your specific goal (e.g., book a hotel room, make a purchase).

The Audience Transparency

Facebook's audience reach is very public, unlike some other ad networks that allow you to target specific groups of people.

Self-targeting allows your business to have a lot of control and transparency over the people you want to reach.

- People who follow you on Facebook.
- Friends of Fans: People who are friends with people who follow you.
- Behaviors or interests: People who fit the criteria you chose based on what they said.

- Remarketing: People who have already been to your site.

While other platforms will automatically place your ads in the best places, on Facebook, you can segment your campaign based on groups of people you already know. This allows you to get more information. Both ways, a campaign could do well.

However, on Facebook, you will be able to see which segments did best, which will allow you to make hypotheses and keep testing and improving your strategies.

Facebook has a lot more power than just demographics to target people. Demographics aren't always a good predictor of how someone lives or what they need to buy.

For example, not all millennials have a lot of debt from college or live in a way that would make sense if they had little money.

Facebook allows you to target people based on a lot of different things about their lives, like their interests, life events, behaviors, or hobbies.

When you use Facebook to market your business, here are the top 10 things that will help your business.

This allows you to not only target with more precision but also make sure that your digital strategy and offline tactics work together so that the same behavioral criteria are used across all of your marketing channels.

Competitor Targeting

Many solutions won't let you go after the people your competitors are going after. People who like other brands can't be targeted on Facebook, so you can't reach them. However, you can still target people who have said they want to buy certain brands. That is based on what the person says, but it could be out of date because it is based on when the person last changed their settings. Still, if used at a large scale, this can be a good way to find people who are already interested in your product.

By making a custom audience of people who are interested in 20+ well-known brands, one can reach a lot of people quickly without having to pay for audience profiles on other channels.

Different Types of Ads

With ten different types of Facebook ads, Facebook is ahead of other social media sites when it comes to the types of ads it makes available. For each stage of the target marketing funnel, there are a few options. Most often, image and video ads are used. It's important to note that almost all ad formats can include some text and images, giving you a lot of room to both describe and show off your business.

Sponsored posts are one ad format to keep an eye on, especially if your business lets other people post on its feed. A user-generated post will get more attention if you put it on your news feed.

Strong user-generated content often outperforms purpose-built ads, because the latter are more easily seen as messages that were made for a specific purpose.

When people make their own content, it is more natural and people are less likely to be defensive when they see it.

Make People Go to Your Site Right Away

As we said earlier, many of Facebook's ad options can help your business get new customers by directing them to your website. It is true that most people who go to Facebook want to stay there and look at the content there.

However, if your ad is interesting and your targeting is very specific, the ad will be enough to get the user to leave Facebook and go to your site.

Measure your Performance

Facebook allows you to report on a lot of different metrics, even though it may seem obvious. This is important to note. Many different metrics are available depending on the type of ad.

These include things that people can do before they leave Facebook (like reach, ad engagements), as well as things that happen outside of Facebook (like a party) (e.g., conversions, revenue). The only thing you need to do is put the Facebook pixel on your site.

Facebook relies on its own tracking pixel to make sure that campaigns are always running at their best. One can use third-party analytics tracking to report basic conversion metrics.

Without it, Facebook can't figure out which user profiles are converting the best and will keep advertising to the same group of people for the whole campaign.

Keep your Existing Audience Interested

One of the best things about advertising on Facebook is that it helps you keep and grow your Facebook followers and fans. While getting referral site traffic, site engagement, and conversions are important, it's also important to keep your Facebook audience happy.

Think of your Facebook profile as your business's second website. It's another way for people to connect with your business online. With your website and any other kind of public persona, over time some people will become less interested in talking to you and talking about your site.

To keep you in their thoughts, they need to be reminded of and given new reasons to keep you in mind. In the same way, users who leave for good need to be replaced with new ones.

Facebook advertising, no matter what your main goal is, always has the important secondary benefit of growing your fan base and giving your current fans something new to think about.

If you have the money, you can run campaigns only for people who already follow you. One does not have to be a follower to see all of the updates. A sponsored campaign that targets followers does not guarantee that all followers will see it. However, it will make it much easier to get closer to people, which will make them want to check your profile more often.

3

BEING A SERVANT VERSUS A SALESPERSON

The key to a successful business, whether it's on Facebook, a fast-food restaurant, or even selling cookies, is to provide value. That value through a product or service must meet the needs of the customer. That is the difference between a successful and a failing business. Your Facebook page is a representation of the value you provide online or offline. Therefore, your page must reflect value through the content posted. As mentioned in the last chapter, you can create your own valuable content or share content from others.

POSTING VALUABLE CONTENT RATHER THAN SELLING

The most common method of providing content is by posting. Facebook offers different types of posts.

Common Types of Posts

- Status posts: These are basic text posts. Good for engaging in conversations and sharing information. Not ideal for driving traffic to your website.
- Photo posts: Great for engagement, and is an attention grabber for potential customers. Especially for product-based businesses.
- Facebook video posts: Much more engagement than regular status posts and pictures.

Now that you have a Facebook Business Page of your own, it's time to start making some content for the page. On Facebook, you can create a lot of different kinds of posts. Each one has its own advantages and can get people excited in different ways, but they all work together. Depending on your brand's social media strategy and goals, you'll likely be making posts that are different from those you've made in the past. Here, we're going over all of the different ways you can post to your Facebook Page, and we'll show you some examples to get you started.

Status Posts

Facebook lets you write a text message and then share it with others (a.k.a. status post). This is the first Facebook post: just text. A few words here and there. None of this. There are no videos. There are no links. They're short and to the point, but if you want to get people to your website or make a sale right away, text posts aren't the best choice. These simple posts also don't get much attention from the social network's algorithm. They don't usually show up in the news feed, and they don't get many views.

There is one thing text posts can do that is good, though: start a conversation. Use a text message to ask a question or ask for feedback, like "What do you think?"

Text posts can also be used to share important information that your audience might be looking for on your Facebook page, like ticket availability or opening hours, which you can do with them.

Photo Posts

Photo posts get a lot more attention than text posts. This is a great way to get a potential customer's attention when they're scrolling through their news feed: show them a picture (or an illustration or an infographic, we're not picky). Especially for businesses that sell goods, photos that show the goods in action can be very effective. These pictures of The Soap Dispensary's bread baskets with tasty sourdough loaves might make carb lovers stop in their tracks, like carb lovers. Breathtaking!

Keep in mind that stock photos come in handy if you aren't an expert photographer, or if you're selling something that is hard to show in a picture. Various sites like Pexels and Pixabay are available on the Internet to obtain stock photos. Some of them offer free photos for commercial use. Take the time to learn about image copyright before you start posting.

Video Posts

Facebook video marketing is a fantastic way to increase user engagement, which will eventually lead to more company success. On a daily basis, people are inundated with messages from a variety of sources, including Facebook, TikTok, Instagram, email, SMS, internet adverts, and many more. In this day and age, it may be difficult to capture and hold someone's attention, even for a short period of time.

It is vital to develop a message that captivates people. When it comes to developing compelling videos, the most important thing is to engage the user. You only get a few seconds to do it.

Video posts get more attention than photos do. Video can be very interesting. For example, Vox posts informative videos directly into its feed so users can watch them right on Facebook and quickly comment, react, and share.

The video will play automatically in the news feed on Facebook, so you're almost certain to get people to watch your post. It's a great way to get them. If you want specific Facebook video marketing advice, here are a few strategies for increasing overall interaction with your Facebook videos.

People are overwhelmed with messages from a variety of sources, including Facebook, TikTok, Instagram, email, SMS, internet adverts, and many more. It is vital to develop a message that captivates people. When it comes to developing compelling Facebook videos, the most important thing is to engage the user. And you just have a few seconds to do that. Due to the short attention span of others, keep videos short, approximately 1 minute. The first 3 seconds of the video are the most important. Give viewers a hook that piques their interests. This makes the video easier to compile and consumes less mobile data. If you must make a longer video, it can lead to more engagement. However, the content must be extremely valuable right from the start to finish.

Ensure that videos are engaging whether they are with or without audio. Add a compelling caption or description to help fill in any gaps, and add subtitles if needed. Tip: adding subtitles will help make your video more accessible for hearing-impaired viewers.

Here are a few strategies for increasing overall interaction with your Facebook videos.

▷ **Posting Videos Directly**

- Adding videos directly to Facebook allows them to appear higher in the News Feed.
- If you want to share a YouTube video on Facebook, skip copying and pasting the YouTube URL and instead publish the video straight to Facebook.
- If you don't like videos that have been made in the past, you might want to try it Live.

▷ **Going Live**

On Facebook Live, you can post a video while recording. Facebook video marketing is a fantastic way to increase user engagement, which will eventually lead to more company success. This is a live-streamed broadcast that you can watch on your Facebook Page, and it's very, very popular. In the spring of 2020, Facebook Live viewership went up by 50%. Live video is a way to connect with your followers in a personal and real way. It is possible to use these broadcasts to answer your customers' questions, show them around, show them how your products or services work, and so on.

Facebook's algorithm has been altered to promote live videos since research shows that live videos retain a user's attention for a longer amount of time.

Live videos have also been found to increase the number of viewers who explore the information on that page (even if it isn't the live video itself).

The theory is that businesses that get in front of their target audience in this direct approach will have a higher degree of relatability and trust. After the recording session is complete, Facebook will capture the live video and upload it immediately to a page.

Ready to start? If you're interested in how Facebook Live can help you market your business, check it out now!

▷ **Creating Facebook Lives**

There are different ways to go Live on Facebook using your mobile device.

◇ **Use the Facebook App:**

- Go to the Page, Group, event, or personal profile that you would like to stream your video from.
- Click 'Live', located at the bottom of your post composer. Write a description.
- Tap Start Live Video to begin the live broadcast.
- Tap finish to end

◇ **Using the Creator Studio Facebook Business Manager App:**

- On the Home or Posts tab, click the compose icon on the top right corner.
- Select the option for Live post.
- Write a description.
- Tap Start Live Video to begin the live broadcast.
- Tap finish to end.

◇ **Use the Webcam on Your Computer to Go Live:**

- At the top of your newsfeed, beneath the "what's on your mind?" status field, click on the Live Video icon.
- You'll be taken to the Live Producer tool, and prompted to choose your video source. Select Use Camera.
- On the left side of the screen, write a description and add an optional title for your live video. Here, you can tag people or places, or choose to raise money with a "Donate" button.
- When you're ready, click the Go Live button on the bottom left of the screen. https://blog.hootsuite.com/facebook-live-video/

▷ **Choose the Right Video Formats**

Facebook accepts a number of formats, although they prefer MP4 or MOV files. MP4 is the most used format for Facebook video advertising.

▷ **Form Matters**

Pay particular attention to whether you choose a square video or a landscape video when arranging your Facebook videos.

Square videos gain more interaction, views, and reach, especially on mobile.

This makes sense given that the majority of Facebook users access the site through mobile devices, and square videos occupy the bulk of real estate on the mobile News Feed.

▷ **Utilize Facebook Insights for Video Information**

Facebook offers statistics such as average watch time, total minutes viewed, and so on. This may give useful information on which videos people find most interesting, as well as aid in the design of future social video efforts.

Analyze your videos through Hootsuite or Facebook Page Manager:

- Is the audience watching my videos from start to finish?
- Are my videos underperforming on any platform?
- Which topics are getting the most views, likes, comments, or shares?
- End with a call to action that convinces your viewers to buy or follow.

Facebook Story Posts

Facebook Stories, like Instagram Stories, are vertical photos or short videos that disappear after 24 hours and can be shared with your friends. Photos show up for five seconds, and videos can be up to 20 seconds long, but they only last that long. There are no rules for Facebook Stories because they live at the top of the News Feed. There are more than half a billion people who see Facebook Stories every day.

Facebook Stories advertising, on the other hand, are likewise 15 seconds long but do not expire after a day. These short video stories may be quite entertaining. They display at the very top of the program, and users are more likely to see them as soon as they open it. Approximately 62% of individuals say they are more interested in a product or brand after seeing a Facebook story.

Pinning Posts

To make sure important news or great content doesn't get lost in the crowd, use this feature. It's your chance to show people why they should like your Page.

You can pin any regular Facebook post to the top of your page so that it doesn't move. As a result, when people visit your Page, this will be the first thing that they see. On the right-hand side, there are three dots. When you're done making the post, you can click on them. You'll be able to "pin post." You will know when a post is pinned it appears at the top of your page and shows "Pinned post" right above it. It will remain there until you decide to "unpin the post" or choose to 'pin' another post.

Linked Posts

Links are posts that share a URL with the people who follow your page. A preview of the website will show up as soon as you paste a link into the box where you're writing.

For example: You share a link to a how-to story from your website. Links can be enticing and eye-catching. You want to stop reading this story right now and click on the link to read or see the rest of it right away!

This means you could share a link to your website, but you could also share content from other sources, like a piece of interesting thought about your products or services. Make sure to add a few words of your own to the post before you hit 'publish.' This will give your readers some context or give them something to take away from the post.

Keep in mind that link posts may get more attention than plain text posts, but far less than photos or videos.

Job Ad Posts

You can also use Facebook Pages to post job listings and special deals if you want to hire people or get people excited about a sale. With a Facebook post, you can also raise money for a good cause. Or you could stare at your post options for hours on end, paralyzed by the choice. It's up to you!

Choose the Right Type Of Post

It takes some trial and error to find the right type of post for your brand. Try out different combinations, and keep an eye on your analytics to see what works best.

Also, you can use social listening to find out how your company makes people happy, and how you can improve. You should keep an eye on what people are saying about your competitors, too! Yes, it's eavesdropping, but it's for business, so it's fine. Listen and learn.

Start by posting content that's already getting a lot of attention on other channels. Have you tweeted something that's getting a lot of attention? To make it even better, share it on Facebook too! Have a page on your blog that gets a lot of new comments all the time? That's another Facebook post that could be good.

In other words, if you want the best results, be sure to share your message with cross-promotion best practices, not just cross-post it.

Let's not forget that posting valuable content is great, but the content is only as valuable as the audience that you market. You will learn later how to target the right audience.

4

GAME, SET, MATCH!

CREATING A STRATEGY FOR POSTING

The first step in developing a winning plan is defining your goals and objectives. You can't quantify performance or ROI until you have objectives.

Define Your Target Audience

Your strategy should aim to collect likes from the followers who have the most potential to bring value to your business through regular engagement, rather than random likes. Do not market your 'steaks' to cattle. What that means is to ensure that the audience that you are marketing to needs and benefits from your products and/or services. Who are your existing customers?

Who is your potential audience? Who are your competitors targeting?

▷ **Learn from Page Insights**

The more you know about your audience, the more material you can develop to meet their demands. Facebook Page Insights makes it simple to collect information about how your fans engage with your Page and the material you provide. Select Insights in the Manage Page menu to view Page Insights. Insights provide information about your Page's overall performance, including demographics and interaction metrics. You may examine stats on your posts to see how many people you're reaching. You'll also see how many comments and responses individual postings receive—information that might help you design future content. The ability to monitor how many people have clicked on your call-to-action button, website, phone number, and address is a major element of Insights. This information is organized by demographics such as age, gender, nation, city, and device, making it easy to tailor future material to your target audience. To get to this information, go to the Manage Page menu and choose Actions on Page.

Identify Customer Goals and Pain Points

Your audience's goals might be personal or professional, depending on the types of products and services you sell. What motivates your customers? What problems or hassles are your potential customers trying to solve? What's holding them back from success? What barriers do they face in reaching their goals?

▷ Asking Your Followers

Customers can also provide social media motivation. What are your target consumers saying online? What can you discover about their desires and needs?

Make a Content Calendar

Sharing amazing material is crucial, but it's also necessary to have a plan in place for when you'll distribute content to achieve the most effect.

Your social media content plan must take into account the time you spend interacting with the audience (although you need to allow for some spontaneous engagement as well, i.e Facebook Live).

- Create a posting schedule: Your social media content calendar should include the days and times that you will publish different sorts of material on each channel. It's the ideal location for organizing all of your social media activity, including photos, link sharing, and re-shares of consumer-generated material, as well as blog articles and videos. It comprises your regular blogging and content for social media initiatives. Your calendar also guarantees that your posts are suitably spread out and published at the ideal times to post. You can arrange your whole content schedule and receive recommendations on the optimal times to post based on your previous engagement rate and impressions.

- If you're beginning from scratch and don't know what sort of material to produce, use the 80-20 rule: 80% of your postings should enlighten, educate, or entertain your audience, while the remaining 20% can explicitly market your company.

- You risk irritating your readers if you post too constantly. However, if you publish seldom, you risk appearing unworthy of following by posting too little. Have a look at the following content idea cheat sheet: https://blog.hootsuite.com/content-idea-cheat-sheet/

- Once you have your social media content calendar planned out, use a scheduling tool to prepare messages in advance rather than updating constantly throughout the day. I use Calendy, however, you can do your research and pick the tool that works best for you.

- When you have gathered enough data, use this information to test different posts, social marketing campaigns, and strategies against each other. Constant testing helps you to learn what works well and what doesn't, allowing you to fine-tune your plan in real-time.

Test Your Strategy With Surveys

Surveys can also be a great way to find out how well your strategy is working. Ask your followers, email list, and website visitors whether you're meeting their needs and expectations, and what they'd like to see more of. Then make sure to deliver on what they tell you.

Keeping Up With Changes

Social media changes every day. Your company will also go through moments of transition. All of this implies that your social media marketing plan should be a dynamic document that you analyze and update as necessary. Refer to it frequently to keep on track, but don't be reluctant to make adjustments to better represent new information.

HOW TO CREATE A FACEBOOK MARKETING PLAN

So you've become an expert in Facebook posts. That was quick! Let's look at some ideas to help you develop a solid Facebook strategy for making the most of your Page.

Identify Your Target Audience

Yes, we keep repeating this because it is crucial to your online success! To effectively engage your audience, you must first understand who you are speaking to. To discover who your target audience is, ask yourself the following questions:

- What is the average age of your target audience?
- Where do they call home?
- What kind of work do they do?
- What are their difficulties?
- What are their Facebook habits and when do they use it?

Of course, understanding the general demographics of Facebook users is also important. After you've gotten a sense of who is using the platform and how that relates to your target customer, you can look at Facebook Audience Insights.

User Demographics From Facebook Audience Insights

The free, built-in Audience Insights tool on Facebook will assist you in drilling down into the nitty-gritty details about potential customers. You can use it to find information on topics such as:

- age
- gender
- education
- relationship status
- language Preference
- usage on Facebook
- previous purchasing activity

After all, if you don't know who you're trying to reach, you're unlikely to succeed.

Establish Objectives

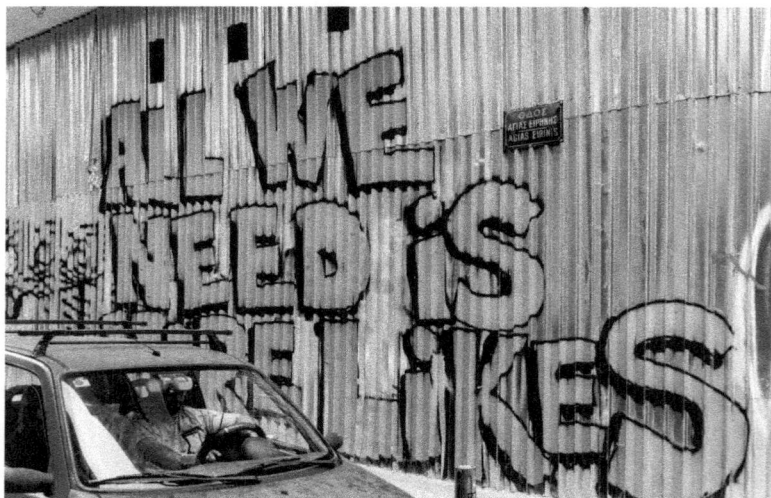

What does success mean for your brand? Sure, it's tempting to view likes as the ultimate measure of success, but if they're not part of a larger marketing strategy, those likes aren't worth much. They're even referred to as vanity metrics at times.

A strong goal that is linked to your business objectives is required for developing an effective Facebook marketing strategy. Every company has different goals, but they should all focus on actions that have an impact on their bottom line.

This could include:

- creating leads
- increasing the number of conversions on your website
- enhancing customer service

We recommend that once you've decided what you want to achieve, you map out specific, measurable steps to get there. We recommend that you use a well-known goal-setting framework, such as SMART goals or the OKR goal framework. More information and inspiring examples can be found in our post on social media goal-setting.

Every post, comment, and ad you create on Facebook should ultimately serve your goals. To stay on track, make a Facebook goal statement for your business, as well as a Facebook-style guide that may inform a consistent look, feel, and voice for all of your material.

(I apologize for assigning so much homework, but successful marketing necessitates a little sweat at times.)

Create a Content Mix

You've identified your target audience. You are aware of your objectives. It's now time to write those posts.

Not every post should extol the virtues of your company. That gets old fast, like a friend's new boyfriend discussing Bitcoin throughout your entire birthday dinner.

Instead, strive to provide value to your followers while also developing relationships with them. Give them content they'll enjoy on a regular basis, and they'll be more open to hearing about your products and services when you bring them up 20% of the time.

The social media rule of thirds is another option for guiding your content mix. A third of your content should involve personal engagement with your audience and the remaining should show your stories, ideas and take home messages to promote your business.

Whatever numerical mix you choose, the goal is to balance promotional material with value.

Facebook penalizes brands that push sales too aggressively. As it turns out, the algorithm isn't a fan of self-promotion. The platform's goal is to prioritize meaningful, engaging content over coupons. After you've decided what to post, the next step is to decide when to post it.

Page Insights can provide some insight into engagement, but our research shows that the best times to post on Facebook on weekdays are at 6:15 a.m. and 12:15 p.m. PST.

Whatever your schedule, remember that it's critical to post on a regular basis.

Create a content calendar to help you balance your content types and organize your frequency. Refer to the free content calendar template earlier in this chapter.

Make Your Page More Engaging

Whatever your Facebook marketing objective is, it will be difficult to achieve if no one knows your Facebook Page exists.

That is why it is critical to a) attract people to your Page in the first place and b) compel them to interact once they arrive.

Expert advice: Pay special attention to your Facebook Page cover image. People will see this as the first thing they see when they visit your Page, so it better look good!

Cross-promotion is a simple way to help people find your Facebook Business Page. Make it easier for people who are already interacting with you on other platforms to find you on Facebook by including a link to your Page in your email signature and newsletter, as well as incorporating Facebook Like and share buttons on your website or blog.

It's a little less scientific to get those views, likes, and follows rolling in: you have to create highly shareable content. Posts that are both informative and entertaining will (hopefully!) encourage followers to share them with their friends.

To increase engagement, keep in mind that you get out of Facebook what you put into it. If you expect your followers to be engaged, you must be engaged as well.

Get chatty because responsiveness is a highly valued attribute of brands. Respond to all messages and comments, answer all questions, and keep the content up to date. (In fact, you should focus on auditing your Facebook Page on a frequent basis to seek for and delete any obsolete information.) Your About section should always be accurate, up to date, and consistent with your brand.

More detailed strategies are available on the Hootsuite website for increasing Facebook Likes, and they also have a list of little-known Facebook tips and tricks.

Think About Using Other Facebook Tools

Once you've mastered the art of managing a Facebook Business Page, there are a plethora of additional ways for brands to find opportunities for engagement beyond posts and comments.

Facebook's Business Manager

It's also a good idea to set up a Facebook Business Manager if you want to take your Facebook Page to the next level. Facebook describes it as "a one-stop shop for managing business tools, assets, and employee access to these assets." How can you possibly resist?!

Simply put, Business Manager is a tool for managing both organic and paid Facebook posts. It also enables you to collaborate effectively with team members as well as outside contractors and agencies.

In our guide to using Facebook Business Manager, we'll walk you through the setup process step by step.

Groups on Facebook

Groups are another excellent "extra credit" tool for increasing engagement. Groups are, in some ways, the online equivalent of your favorite coffee shop or community center. They are online communities where people can share information and ideas...and hopefully, a love for your brand. With 1.4 billion people using Facebook Groups each month, it's an audience that can't be overlooked.

You can also use Facebook Groups to demonstrate your expertise and provide added value to your fans by providing exclusive content or special deals to 'members.' This is an excellent method for fostering trust and long-term loyalty.

If you're ready to add this feature to your Facebook Business Page toolset, we've got step-by-step instructions for creating your own Facebook Group right now.

If people are enthusiastic about your business, you may not even need to establish a unique location for them to congregate: Occasionally followers may start their own Facebook Group devoted to your company. If you come across such Groups in the world, it's a good idea to join so you can monitor the debate and clarify any misunderstandings or hate.

In general, however, fan-created Facebook Groups are an excellent indicator that you're doing something right. You're very fortunate!

ChatBot for Facebook Messenger

Every month, Facebook users send 20 billion messages to businesses. Twenty billion messages! If you are not available for conversation with your customers via this platform, you may be passing up a chance to connect.

It's not enough to simply have Facebook Messenger installed. It all comes down to being extremely responsive to customer messages. According to Facebook research, users expect a business to respond almost immediately. One Facebook user poll stated that he would only wait 10 minutes for a response before moving on to another brand.

What is the answer to these expectations? Chatbots are automated response tools that are always available to assist a potential customer, such as the one on Booking.com.

Use Facebook Ads and the Facebook Pixel

So you wrote the perfect post: the wording is perfect, the imagery is stunning, and the question is intriguing. Here's hoping your followers notice it.

That's right: not everything you post on your Facebook Page will appear in the news feeds of your followers. You might be surprised by the number of followers your organic posts will most likely attract:

- 8.18% organic reach for Pages with fewer than 10,000 followers.
- 2.59% organic reach for Pages with over 10,000 followers.

The Facebook algorithm, for better or worse, prioritizes posts from users' friends and family. This means that businesses and brands can't always stand out from the crowd.

Sometimes your excellent content just needs a little nudge. Fortunately, Facebook Ads allow you to broaden your reach without breaking the bank.

A Facebook ad, like traditional advertising, is content that you pay to share with a specific, targeted audience. Whether you want to increase brand awareness, engagement, or traffic, the goal is to get your brand in front of the right people.

Check out our complete guide to Facebook advertising for detailed instructions on how to launch a targeted Facebook ad campaign.

Even if you aren't ready to dive into Facebook Ads, it's a good idea to set yourself up (for free!) with a Facebook pixel right now.

A Facebook pixel is a small piece of code that you can add to your website to track Facebook conversions, remarket to people who have already visited your website, and create targeted custom audiences for future ads.

The pixel will begin collecting data as soon as you place it on your website. As a result, whenever you're ready to advertise, you'll have powerful data for retargeting campaigns at your fingertips.

How to Measure the Success of Your Strategy

Facebook marketing requires ongoing attention; it is not a set-it-and-forget-it condition. Tracking and measuring are critical for understanding what ended up working and what didn't. This way, you can constantly learn, tweak, and try again to improve your strategy. You can directly track audience engagement using Facebook Insights, which measures metrics like...

- reach for likes (how many people saw your posts)
- participation (how many people liked, clicked, shared, or commented on your content)
- which of your posts causes people to unlike your Page

Facebook Insights will assist you in determining which types of posts work best for your Page, allowing you to determine whether your current content mix is effective. Check out our beginner's guide to Facebook Analytics for more information.

Purchases and other website conversions must be tracked outside of Facebook using external tools such as Google Analytics, Hootsuite Impact, UTM parameters, and Hootsuite Social Advertising.

Tracking your progress entails more than just counting your victories and congratulating yourself on your accomplishments. It's also about keeping track of what's not working. Don't be overwhelmed! You can compare organic and paid content using an integrated tool like Hootsuite Social Advertising. With a unified view of your Facebook and other social media activities you can make informed choices for future online campaigns. For example, if a Facebook ad is performing well, you can adjust ad spend across other platforms to support it. Similarly, if a campaign is underperforming, you can pause it and redistribute the budget without leaving your Hootsuite dashboard.

More About Hootsuite Advertising

Phew! We know what you're thinking: there's a lot to learn about Facebook marketing. But the good news is that you don't have to spend any money to get started.

So go ahead. Get your hands dirty, and learn as you go. When you're ready to take things to the next level, more complex strategies and campaigns will be available... and our dozens upon dozens of resources and guides will be there to assist you every step of the way.

Hootsuite can help you manage your Facebook presence alongside your other social media channels. You can schedule posts, share videos, engage your audience, and track the effectiveness of your efforts all from a single dashboard.

5

ADVERTISE LIKE A BIG BUSINESS WITHOUT BIG COST

FACEBOOK AD AND LIKE CAMPAIGNS

The goal of Facebook ads is to get your message in front of the right people at the right moment. Individuals who are most likely to invest in your goods or services.

Before you get in, it's critical to understand the various Facebook ad kinds and targeting choices if you want to receive the best results.

Types of Facebook Ads

▷ **Image Ads**

These straightforward advertisements are an excellent way to get started with Facebook paid advertising. You may make one in a matter of minutes by boosting an existing post with a picture from your Facebook Page. Image advertising might be basic, but they don't have to be dull.

▷ Video Ads

Video advertising may display in the News Feed and Stories, or as in-stream ads in bigger Facebook videos. Video advertisements may show your team, products, or services in action. Video commercials, like picture ads, may take a much more creative approach. You don't have to utilize recorded video content in your video commercials. You may also use GIF-like images or other animations to draw attention to your deal or to explain it.

▷ Lead Ads

Lead advertisements are only accessible on mobile devices. This is because they are expressly intended to allow individuals to provide you with their contact information without having to type it in.

They're ideal for gathering newsletter subscriptions, registering people up for a product trial, or letting users request additional information from you. Several manufacturers have utilized them effectively to persuade people to test drive automobiles.

Utilizing Facebook lead advertisements is a terrific method to fuel your sales funnel. It explains all you need to know to make the most of this critical form of Facebook ad campaigns.

▷ Messenger Ads

Facebook Messenger advertisements allow you to reach the 1.3 billion individuals who use Messenger on a monthly basis. Simply choose Messenger as the preferred location when designing your ad. You must also pick your Facebook feed.

In the Facebook feed, you can also place "click-to-Messenger" advertising. These advertisements have a call-to-action button that initiates a Messenger chat with your Facebook Page. It may be used to conduct a one-on-one chat with one of your salesmen or customer care representatives.

How to Create Successful Ads

You're browsing through your page when you stumble upon an amateur Facebook ad design. The picture is hazy, the language is illegible, and you have no clue what they're promoting.

Getting started with Facebook ad design is not simple. There is a great deal to think about. It's no wonder that when it comes time to create their next commercial, companies and marketers alike are at a loss for ideas.

Ad design is significant because it is how we visually convey and show our businesses to the general audience. You may have a fantastic product and a fantastic deal, but it will be difficult to get momentum if your ad is aesthetically unattractive, confused, or just dull.

Excellent ad design draws your readers' attention, educates them on who you are and what you are giving, and instructs them on what they should do to take that all-important next step toward conversion.

Each of these tips is essential for both seasoned professionals and newcomers. They will prevent you from spending money on advertisements that are ignored by readers and fail to convert your target demographic.

If you're ready, let's get started on discovering the secrets of professional ad design.

▷ Create a Clear Call to Action (CTA)

Every campaign or ad style in the world can be divided into two categories: advertising that engages your prospect's attention and ads that drive a direct action such as a sale, app download, or lead.

In an ideal world, your campaign is doing both. However, in most circumstances, you'll receive only one or the other. Brand recognition is important. It's a wise technique that will help your company grow in the long run. However, far too many efforts attempt to combine brand recognition with direct reaction. It seldom works unless you're a marketing genius.

As a result, CTAs connected to content consumption, such as following your Facebook page, subscribing for additional material, or collecting email subscribers, are more suited for innovative brand awareness efforts. And, rather than attempting to engage or amuse, direct response commercials are better suited to addressing typical purchasing objections.

When creating your adverts, keep in mind that viewers will only give your ad a few seconds to wow them. Make the most of your time by placing your value proposition front and center while making your call to action clear and effective.

▷ **Use a Look-alike Audience**

We have briefly talked about look-alike audiences before. Look-alike audiences are effective because they allow you to target comparable prospects on Facebook using existing data (for example, those who bought a product from your website). This provides you with a great foundation for testing and improving your audience targeting. Now we need to find out how you can create a Facebook look-alike audience. Follow these procedures with your preferred Facebook ad tool.

◇ **Navigate to Your Advertising Manager's Audience Section**

1. Create a look-alike audience by clicking the Create a Look-alike Audience button.
2. Select Create Custom Audience, followed by Customer File.
3. You may then import an Excel file of clients, such as your email list or a PayPal customer list.
4. Choose a nation where you'd want to meet others who share your interests.
5. Using the slider, choose the size of your target audience.
6. Click the Create Audience button.

If you want to target the most prospective lead prospects, establish look-alike audiences that target one to two percent of a country's population rather than aiming for ten percent. Don't forget to exclude custom audiences of those who have already converted for the best results. Later, refine with more precise targeting. After you've launched your first campaign, you may fine-tune your audience targeting approach by making the changes shown below. Add them one by one to see whether they have an effect. Hootsuite's AdEspresso blog describes how Facebook targeting works.

Under behaviors, you may target particular device owners, such as those celebrating an anniversary in the next two years or those who have just made a business purchase.

Another strategy is to begin by testing broad audiences and then add more specificity as you go, resulting in a more focused and higher converting audience each time.

▷ **Create an Attention-Grabbing Headline**

Although Facebook's targeting features make it simpler to get your advertising in front of the individuals who matter most to your campaign, it is completely up to you to persuade them. To accomplish this, you must pay great attention to every detail.

The headline of a Facebook Ad may seem to be a little element of the jigsaw, but this couldn't be further from the truth: The power of your headline will surely affect a user's choice to take action on your Facebook Ad. After all, it establishes the tone and expectation for all that follows.

That being said, you want it to be memorable, eye-catching, and, most importantly, clear. We've prepared some criteria and recommendations to help you create a Facebook Ad title that does all of the above.

▷ The Building Blocks of a Clickable Facebook Ad Headline

Below you will find three guidelines to help you create headlines that grab the attention of your target audience.

▷ Customize the Title To Make Your Ad More Personal and Interesting

The title, which appears directly below your featured picture, is often the final thing Facebook visitors read in order to comprehend precisely what you're giving. In other words, a lack of a headline leads to an overall lack of clarity.

For example an ad headline stating: "Subscribe to our magazine at only $12 for 12 weeks" may be personalized, but it lacks a convincing justification or a clear advantage for subscribing. A more intriguing title may be, "Subscribe for $1 a Week" or "Award-Winning Reading for Only $1 a Week."

◇ Be Straightforward

With over 3 million posts published on Facebook every minute, you must aim for a level of headline clarity that is not easily overlooked.

The headline should clearly state the "what" (for example specific product) and "why" (for example $50 discount). Clearly describe what people are being encouraged to sign up for.

◇ **Short and To-The-Point**

To reiterate the preceding advice, make your Facebook Ad title brief, sweet, and to the point. To optimize interaction, Facebook suggests that Ad headlines be between 25-40 characters long, thus your attention should be on generating something that is simply comprehended and value-oriented.

Remember in most cases "less is more": and don't underestimate the effectiveness of simple and sweet Facebook Ad headlines.

Do you need assistance producing more precise content? To get in the zone before composing your headline, practice writing in short form (tweets, six-word memoirs, etc.). Ask a question, inspire curiosity, provide a benefit, or give a command.

▷ Use a Video or Image That Clicks With the Headline

Make sure your headline works in collaboration with your visual graphics. Take into account the emotional offering of your headline and video or picture. To create a successful headline focus on the what and why of your ad.

▷ The Body Should Put Customers/Prospects at Ease Before Performing the CTA

The CTA should have a meaningful hook that will entice your prospect to click on it. Your CTA should be action-oriented and provide enough value to your prospect. It should be equally enticing for the prospect to believe you and act. Your ad content should clearly communicate why it's valuable for people to follow your CTA and share their information with your business.

CREATING YOUR ADVERTISEMENT

Step 1: Select Your Objective

Log in to Facebook Ads Manager, go to the Campaigns tab, and then click Create to begin creating a new Facebook ad campaign. Based on what you want your ad to accomplish, Facebook provides 11 marketing objectives. Here's how they relate to business objectives:

- Brand awareness: Make your brand known to a new audience.
- Reach: Get your ad in front of as many people as possible in your target audience.
- Traffic: Direct visitors to a specific website, app, or Facebook Messenger conversation.

- Engagement: Reach a large audience to increase the number of post engagements or Page followers, increase event attendance, or encourage people to take advantage of a special offer.
- App installs: Persuade people to download your app.
- Increase the number of people who watch your videos.
- Lead generation is the process of bringing new prospects into your sales funnel.
- Messages: Encourage customers to contact your company via Messenger.
- Conversions: Encourage people to take a specific action on your website (such as subscribing to your list or purchasing your product), through your app, or through Facebook Messenger.
- Catalog sales: Link your Facebook ads to your product catalog to show people ads for products they are most likely to buy.
- Store traffic: Drive customers to nearby brick-and-mortar stores.

Choose a campaign objective based on your objectives for this specific ad. Keep in mind that while conversion-oriented objectives (such as sales) can be paid for per action, exposure objectives (such as traffic and views) will be paid for impressions.

If your objective is engagement, you must specify the type of engagement you desire. Some of the options you see in the following steps will change depending on which objective you select.

Select "Create New Campaign" in Facebook Ads Manager and click 'Continue'.

Step 2: Give Your Campaign a Name

- Name your Facebook ad campaign and specify whether your ad falls under any special categories, such as credit or politics.
- If you want to run an A/B split test, go to the A/B Test section and click Get Started to make this ad your control. After the ad is published, you can select different versions to run against it.
- Scroll down a little further to select whether or not to enable budget optimization. This option is useful if you use multiple ad sets, but you can leave it turned off for the time being.
- Choose a campaign name and specifics. Then, click the 'Next' button.

Step 3: Create a Budget and a Timetable

You will name your ad set and select which Page to promote at the top of this screen. The next step is to determine how much money you want to spend on your Facebook ad campaign. You have the option of setting a daily or lifetime budget. Then, if you want to schedule your ad in the future, enter the start and end dates, or choose to make it live immediately.

Keep in mind that scheduling your Facebook paid ads may be the most efficient way to spend your budget because you can choose to only serve your ad when your target audience is most likely to be on Facebook. Only after you've created a lifetime budget for your ad can you set a schedule.

Step 4: Set Up Your Audience

Scroll down to begin constructing your ad's target audience. The first option is to create a custom audience of people who have previously interacted with your business on or off Facebook. We have a separate guide to walk you through Facebook custom audiences, so we'll concentrate on the targeting options here.

Begin by deciding on your target location, age, gender, and language. It's worth noting that you can specify whether or not cities of a certain size should be included or excluded under location.

As you make your choices, keep an eye on the audience size indicator on the right side of the screen, which gives you an idea of how far your ad could reach.

You'll also notice an estimate of the number of Page likes. Because Facebook has more data to work with, these estimates will be more accurate if you have previously run campaigns. Remember that these are only estimates, not guarantees.

It's now time for the fine-grained targeting. Remember that effective targeting is critical to maximizing ROI, and there are numerous ways to target your audience with Facebook Ads Manager. You have two fields here to make your audience as specific as you want it to be:

- Detailed Targeting: Use this option to include or exclude persons based on their demographics, hobbies, and actions. You have a lot of leeway here. You may, for example, opt to target those who are interested in both meditation and yoga while eliminating those who are interested in hot yoga.
- Connections: You may target or exclude users who are already linked to your Facebook Page, app, or event. If you wish to reach a fresh audience, for example, you might pick "Exclude those who like your Page." If you wish to advertise an offer or new product to existing admirers, choose "People who like your Page." You may also opt to target those who have already engaged with your brand's friends.

Step 5: Decide on Your Facebook Ad Placements

Scroll down to select the location of your ads. If you're new to Facebook advertising, the most straightforward option is to use Automatic Placements. When you choose this option, Facebook will automatically place your ads on Facebook, Instagram, Messenger, and the Audience Network where they are most likely to perform well.

Once you've gained some experience, you might want to target your Facebook ads to specific locations. Your options will vary depending on the goal of your campaign, but they may include the following:

- Type of device: mobile, desktop, or both.
- Facebook, Instagram, Audience Network, and/or Messenger are examples of platforms.
- Messaging, Feeds, Stories, in-stream (for movies), search, in-article, applications, and sites are all options (external to Facebook).
- iOS, Android, feature phones, or all devices are examples of specific mobile devices and operating systems.

Step 6: Establish Brand Safety and Cost Controls

Scroll down to the Brand Safety section to exclude any content that would be inappropriate for your ad to appear.

You can, for example, choose to avoid sensitive content and add specific blocklists. Specific websites, videos, and publishers can be excluded from blocklists.

Finally, you can fine-tune your ad bidding strategy and bidding type, as well as add an optional bid control. If you're new to Facebook advertising, you can begin by using the default settings.

If you have more experience, customize the options here to best match your budget strategy with your campaign objectives.

When you're satisfied with all of your options, take one last look at the estimated reach and conversion rates. Click Next if you're satisfied with what you see.

Step 7: Create Your Advertisement

Choose your ad format first, then enter your ad's text and media components. The formats available will vary depending on the campaign objective you chose at the start of this process. If you're working with an image, try turning it into a video by clicking Turn into Video. Alternatively, click Create Slideshow to create a Slideshow ad using the built-in Video Creation Kit.

Use the preview tool on the right side of the page to ensure that your ad looks good in all possible placements. When you're satisfied with your selections, click the green Publish button to begin running your ad.

Important Ad Objectives:

- Timing
- Ad placement
- Target audience
- Impressions
- Reach
- Lead generation
- Conversion
- Link clicks

Facebook Ad Text and Objective Specifications

Keep the recommended character counts in mind when creating Facebook ads. Anything that extends outside these text limits will be snipped. You must also recognize which types of Facebook ads are compatible with each of these ad campaign objectives.

▷ **Ads with Images or Video**

- 40 characters for the headline
- 30 character link description
- 125 characters for the body text

▷ **Advertisements on Facebook Stories**

There is no set character count for text. Aim for 250 pixels of text-free space at the top and bottom of the ad.

▷ **Advertisements in the Messenger Inbox**

- 40 characters for the headline
- Description of the link: n/a
- 125 characters for the body text

How Much Does Facebook Advertising Cost?

It depends, is the answer. According to AdEspresso's extensive research, the primary factors influencing Facebook ad costs are:

- Timing: the month, day, and even the hour can all have an impact on ad cost.
- Bidding strategy: if you go for the lowest price or a set bid cap.
- Ad placement: higher-competition positions are more expensive.
- Ad relevance: low engagement, quality, or conversion rankings for your ad can increase costs.
- Target audience: audiences with a higher level of competitiveness are more expensive.

Keeping this in mind, AdEspresso computed the average cost per click for Facebook advertising in the third quarter of 2020. The average costs, broken down by campaign strategy, were:

- Impressions: $0.98
- Reach: $1.03
- Lead generation costs $0.67
- Conversions: $0.25
- Link clicks cost $0.16.

Ad Rejection: Reasons and Actions to Take To Avoid Rejection

Using Facebook advertising implies that you will encounter ads that are not authorized for one reason or the other. It's not the end of the world, but it may be perplexing for those of us who sincerely want to do the right thing. If you've ever had an ad rejected with a vague explanation, you know how tough it may be to determine which element of your ad caused the issue.

Worse, it might often take more than 24 hours for the ad to be assessed in the first place. If you are refused, you may have to wait another 24 hours. Let's go through some of the most prevalent reasons for ad rejections that many people aren't aware of, and how to prevent them. The idea is to assist you in getting more advertisements accepted and remaining on Facebook's winning side.

The first point to remember is that Facebook's users are more important than its advertising. If you and I both quit advertising on Facebook tomorrow, it wouldn't matter since another two marketers would rapidly fill the void. There will always be new sponsors as long as there are more than 1 billion monthly active users.

However, if the users left, there would be no one to promote to, therefore no one would continue to run advertising. This will have a significant impact on Facebook's bottom line.

As a result, it stands to reason that Facebook's top objective is to safeguard its users' experiences. There must be a balance between marketers' ability to sell on the network and the experience of Facebook's users. Which is why ad policies exist. Advertisers would very certainly damage the platform if they didn't have them.

Keep the user experience in mind next time you're designing a Facebook ad campaign. Developing this mentality will not only help you get much more Facebook advertisements accepted, but it will also enhance your conversion rate.

Deceptive advertising may get people to click, but it's a short game that will cost you in the long term. You're probably irritating ten people for every one that clicks. And for the other ten, you've harmed your brand, maybe to the point that they'll never become your clients.

And for every individual who clicks and doesn't receive what they anticipate or feels deceived into clicking, you've permanently turned people off from your brand.

Let's look at a list of some of the most typical reasons why your Facebook ad may not be accepted and what you could do instead.

- Prohibited content not allowed: Advertising about weapons, tobacco, drugs (including Cannabis and CBD), violent images, sexual content, illegal products, or services and copyright material / content. Misinformation is prohibited.

- No before and after images or images that contain improbable results: Before and after photos are not permitted on Facebook. If you want to promote health products, your content must not imply or attempt to generate negative self-perception.

- No deceptive claims: You are not permitted to make false claims. This entails making claims that aren't likely to apply to all readers. For example, a sweeping statement like: "Make $5,000 in 7 days." You can't make this claim unless everyone can achieve it.

- No advertising of Multi-level marketing schemes: MLM's are not permitted to be promoted on Facebook. They will instantly dismiss these ads, and you risk having your ad account banned.

- If you mention Facebook, follow the brand guidelines: If you mention Facebook or any of their products or services, you should adhere to their branding guidelines, otherwise your ad will be rejected.

- Low-quality or disruptive content not allowed: You should also make certain that your landing pages do not provide an unexpected or disruptive experience. This includes sensationalized headlines and irrelevant content.

- Landing pages with little original content: Make sure your landing page isn't too thin/sparse and has some original content. Landing pages with little content may result in ad rejection.

- Personal attributes cannot be mentioned: You may not mention, assert, or imply personal attributes in your ads. This includes factors such as race, ethnicity, religion, beliefs, age, sexual orientation, disability, and medical conditions, among others.

- It does not make users feel bad about themselves or highlight imperfections: Facebook wishes to provide a pleasant user experience. Advertisements are not allowed to point out flaws or make users feel inadequate or bad about themselves.

- Profanity is not permitted in advertisements. The most common reason I see ads get rejected is when a page post that contains profanity is boosted or converted into an ad.

- Your URL description must match the domain to which your ad links, or it will be rejected immediately. Unless you have a compelling reason to change it, leave the URL Description field blank.

- Dissuasion against vaccines: Vaccine advertisements cannot discourage the use of vaccines.
- There are no dating apps or services allowed: This is restricted and can only be done with Facebook's prior written permission.

If the ad is denied, make changes as quickly as possible (Facebook will provide you with the rejection reason) or file an appeal with Facebook if you feel the ad was rejected in error (Hubbard).

Ad Performance Monitoring and Optimization

In the Facebook Ads Manager dashboard, keep a close eye on how your ads are performing. If a campaign isn't going well, invest in a high-performing ad instead. If you're just getting started, it can be a good idea to run a few commercials with limited audiences and costs. Use the winning ad as your primary campaign once you've decided what works best. There are technologies available to assist simplify—and perhaps automate—this process.

Hootsuite Social Advertising is a platform for managing both organic and sponsored content. You can quickly retrieve actionable information and create custom reports from the dashboard to demonstrate the ROI of all your social initiatives.

With a comprehensive view of all social media activity, you can simply make data-driven changes to ongoing initiatives (and get the most out of your budget). For example, if a Facebook ad is doing well, you may alter ad expenditure across other platforms to support it. Similarly, if a campaign is failing, you may halt it and reassign the funds.

Metrics

Depending on your social media strategy, you can use several metrics to assess your ad success. Testing is the most efficient technique to verify that your advertisements are doing as well as they possibly can. Your budget will heavily influence the sort of testing you conduct.

For tighter budgets, larger tests are more beneficial:

- Placements
- Targeting
- Media types and Creatives

Facebook makes it simple to do A/B testing and provides a useful report once the test is over to let you know which ad won. However, split testing is not the compulsory testing type. You may also test your Facebook advertising by creating a conventional campaign or a dynamic ad.

Do you have a bigger advertising budget? You may do granular testing to determine which color backdrop works best for your ad image (among other things).

▷ A/B Testing

To conduct A/B testing, you divide your audience into two equal groups. The same ad is then broadcasted to each group. Then you compare the results to see which one works best for you.

Always modify only one aspect in both variations while undertaking social testing. You're evaluating their reaction to the whole ad. If you change the image and headline, you won't know which is causing the difference in response. Multiple tests are required to test a large number of components. Remember: compare responses and choose a winner.

An A/B test on social media looks like this:

- Select a test element.
- Examine current information for best practices, but don't be hesitant to dispute assumptions.

- Make two versions based on your study (or gut). Remember to only change one piece between versions.
- Demonstrate each version to a subset of your group.
- Keep track of your results.
- Pick the winning variant.
- To improve your findings, share the winning variation with your full list, or test it against another tiny variation.
- Share what you discover to establish a brand library of best practices.
- Restart the process.

Metrics to take into account:

- Cost Per Thousand Impressions (CPM)
- Post Reach
- Click-Through Rate (CTR)
- Conversation Rate
- Conversion Rate
- Cost-Per-Click (CPC)

Consider Boosting Posts

Facebook's organic reach is shrinking. Its algorithm is updated on a regular basis. Facebook has shifted its focus to provide a more organic and enjoyable experience to users by limiting the reach of businesses and brands. So, if you're a marketer unable to get results, it's usually not your fault. Unless you have thousands of followers, your message will not reach as many people as you would like. Because you don't require Ads Manager experience to use Boost posts, it's a simple approach to advertise on Facebook. Simply select a post from your page's timeline and press the Boost Post button. Boosting a Facebook post can help your content reach more people, even individuals who don't already like or follow your page but are interested in your offering.

In certain cases, boosting a Facebook post is the best option. For example:

- Feature update: If you've created a new feature to your Facebook page, for example, a Shop section, boosting might be a wonderful approach to inform your fans about it.
- Posting new material: Facebook Boost Post is an excellent tool for promoting new content. Most marketers enhance Facebook posts whenever they share useful content in order to reach a larger number of their followers. While it is preferable to boost a post that has a high organic reach, there are some posts that you know would receive higher interaction if boosted.
- Build brand awareness: If you don't need conversions but only want to raise awareness, boosting is a better and less expensive choice than Facebook advertisements. Because individuals are influenced by the thoughts and behaviors of others, increasing participation can give crucial social evidence. You can create brand recognition and establish an online community this way.
- One-time events: Events may not always necessitate full-fledged campaigns. With just a little help, you can spread the word and encourage more people to attend the event. This is also true for special deals and

promotions. For example, PanIQ Escape Rooms, a prominent escape room provider, utilizes Facebook Boost Post to publicize discounts and special deals.

Before you give a Facebook post a boost, think about the following:

- Is there a purpose to this post?
- Is there a clear call to action in the post?
- If so, does it lead to a useful landing page?

The post may not be worth the effort if you can't answer all of these questions with 'yes.' Don't just boost every post. If you're convinced that the boosted post will help your readers get closer to your end goal for the post, then go ahead and do so.

How to Create a Facebook Likes Ad

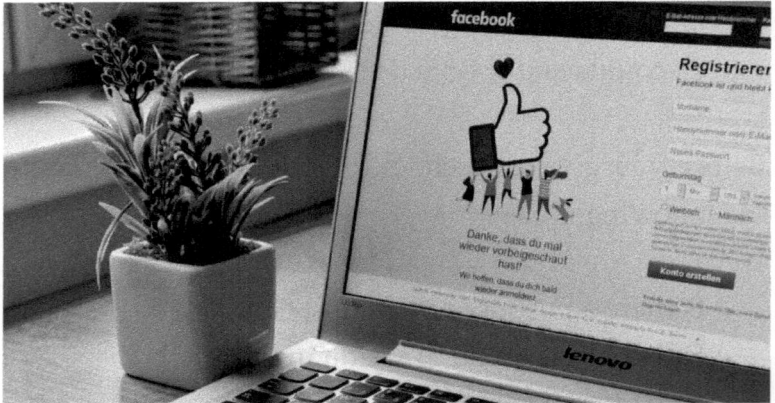

Having a large number of likes on your Facebook Page is the simplest form of social validation. People will be more interested in your brand if you have dozens of Facebook likes. More Facebook likes = increased trust, which equals increased sales (Adespresso, 2021).

Page Like advertising on Facebook are advertisements for corporate pages that are intended to boost page likes and followers for the promoted page. These advertisements are simple to develop and can be launched straight from your Facebook page or by launching a new campaign in the Ads Manager and choosing "Page Likes" as the engagement campaign target.

Follow These Easy Steps:

1. Visit your Facebook page.
2. Click on 'Promote.' At the very top of your page, choose "Create new ad."
3. Under the Goals tab, click the See All button to see all of the options.
4. As your ad target, select the option to increase page likes.
5. Fill out your ad's information. We'll make some suggestions based on your Page, but you're free to adjust them as you see fit. Your ad should include these sections:

- Ad creative: Text plus a picture or video are required.
- Target audience: can be selected from a pre-existing list or created from scratch depending on a variety of criteria.
- Budget: Choose from one of the suggested daily budgets, or enter your own.
- Duration: If you don't see a time frame that works for you, enter a date range.
- Method for payment: You may want to check your payment options. Changing or updating your payment method is possible if necessary.

6. Click "Promote now" when you're done.

DRIVING TRAFFIC FROM YOUR FACEBOOK AD TO YOUR WEBSITE OR LANDING PAGE

W hen you advertise on Facebook using pay-per-click (PPC), you'll need a landing page to help you convert visitors. These pages stand out from the rest of your website since they go hand in hand with your Facebook advertising. Typically, the pages on your website include details about your company. A landing page created exclusively for Facebook adver-tisements, on the other hand, encourages Facebook users to register for or claim an offer tailored just for them, such as a free trial or special offer.

HOW AND WHY TO CREATE A LANDING PAGE

Etsy's Facebook ad campaign, for example, features a prominent call-to-action button that encourages viewers to click through and discover more. For additional information, users can click on the icon and go to a special landing page on Facebook.

Facebook will also help to drive traffic to your website. Consider hiring a professional to create a website for your business. Some online companies already have websites that affiliates can leverage and drive traffic to through example click funnels.

By adding the URL to your Facebook page, you may connect your landing page to Facebook. When creating a Facebook ad, put the URL of your landing page to the website link field in the Ads Creative. A simple Facebook landing page may be created for free using Seed-Prod. At no additional expense, you may generate leads using this tool's pre-built landing page design.

Turning Traffic to Leads and Lead Conversion Using Landing Pages

Surely, you've thought about our hottest audience and how their understanding of your product or service may be reflected in your ad. It is the goal of a landing page to gain information about your website's visitors (usually name and email). By doing this, you'll be able to compile a list of potential customers to whom you may send future marketing messages. One of the most difficult problems for business owners to overcome is the lack of an email list. When visitors come to their site, they don't have a mechanism to follow up and close the deal. As outlined in the Sales Funnel model, your lead magnet should help your potential client solve a specific problem in return for their email address. Just by entering their email address, users are rewarded with time-saving benefits.

Consider a personal trainer who wishes to increase the number of clients he or she has. The end result is that a client pays a big monthly fee to get in shape with their help. A little portion of the final product may be offered as a lead magnet, and their traffic would receive free, useful information that aids them in their road to fitness. The trainer might use a certain workout as an LM to assist her key customers to lose abdominal fat. Customers may download this useful information and sign up for her email list to get updates. Following up with the potential trainees, she may provide them with further information and eventually turn them into full-time clients through one-on-one sessions.

Product splintering refers to the practice of dividing the final product into smaller pieces. Potential clients may not be able to solve their whole problem, but the lead magnet allows them to solve a little one for free and gets them to your firm. Do you know what your ideal clients are willing to pay you for? In a lead magnet, you might offer a small chunk of that product or service that you could break off and give away for free.

Use Lead and Conversion Ads

Facebook lead generation advertising allows you to capture people's contact information directly on Facebook in order to produce qualified leads. Previously, if you wanted to gather people's information, you had to send them to a website landing page and have them fill out a form there.

People are sometimes hesitant to fill out forms, especially on mobile devices where they are difficult to use. The main idea behind Facebook's lead generation ads is to make the form as simple as possible for the user—specifically it's designed for mobile users (the ad only appears on the mobile app), they can complete it without leaving the app, and many of the fields are automatically populated with information from Facebook. So they're given with a form that's already half done and extremely simple to use, which significantly decreases friction and generally leads to more completions.

Facebook lead generation advertisements are appropriate for any scenario in which a user fills out a form, however, the following are some popular examples:

- requesting a product or service quote
- requesting a product or service demonstration
- subscribing to a newsletter
- downloading restricted content
- registering for an event
- obtaining a product sample

- entering your postcode to locate a store near you

The advertisement is divided into three sections: a 'sponsored' advertisement that seems to be an organic post, the form itself, and a "thank you" message that shows when the form is finished. The sponsored advertisement is similar to other forms of Facebook advertising in that it appears as a regular post with text, an image or video, and a call-to-action (CTA) button. When a sponsored ad shows in the news feeds, the Instant Form appears when the user clicks on the CTA. This may be tailored to your specific needs, with choices for multiple-choice questions, conditional replies, time and data pickers, and more. It may also include a pre-page that provides the user with further information or encourages them to finish the form. Finally, the material you give might be dynamic, based on the sites and groups liked by the prospect on Facebook. When paired with Facebook's exact targeting, the advertisement becomes even more accurate and impactful.

If the advertisement and form are appealing enough to the user, and they joyfully exchange their personal information, the last portion displays a "thank you" message in which you may express your thanks for their interest and advise them of the following stages. It may also include another CTA button that directs the prospect to your website or another promotional page, as well as a "call us" button that allows them to call you right away if they so choose.

When the form is submitted, the prospect's information can be sent to one of many destinations. If you use a CRM that is linked to Facebook, such as HubSpot, the information is sent there and can be recognized as a qualified lead. If you do things manually, you can find the data in Facebook's Ad Manager under the "Lead Center" which contains every lead you've created thus far. After that, leads may be downloaded as a spreadsheet and used by your sales and marketing teams. When individuals fill out the form, Facebook calculates and charges a fee per lead.

7

WHY AM I NOT GENERATING SALES?

There is certainly no worse feeling than investing the time, money, and energy crafting content just to put it out there and discover that you aren't receiving the response you want, or, more specifically, that you aren't generating new leads. It's difficult to put in all of the work to create content when you're not seeing any real returns.

It's simple to justify abandoning content marketing. Unfortunately, leaving the scene won't work in today's day and age. While it may seem like the easier option, you should first examine why you aren't obtaining the desired outcomes and determine whether any changes need to be made.

REASON WHY ADS ARE NOT CONVERTING AND SOLUTIONS

Skipping Steps of Your Sales Funnel

To establish a complete functional procedure that leads to the desired result. A working funnel is better than a perfect one. You will produce a confusing mess if you do not follow the funnel development process from top to bottom. Jumping the funnel is based on erroneous data. The system will likely fail.

To start, you need to get your prospect's attention with your post or type of Facebook ad. Conversion is the final phase. Conversion occurs when prospects phone, email, download, or purchase your product or service. Everything else in the funnel is dependent on how you convert a prospect into a customer.

This is a testing process. But first, create a plan that outlines your entire sales or marketing funnel. Once the plan is set, test each step sequentially.

Begin from the top and work your way down. The only constant in the funnel is your product or service. During the funnel creation process, only change the parts within the funnel you are working on. For example, you can test Facebook ads with different options, but don't change your 1,379 landing pages.

Only test one element per time as you refine your funnel. If you test more than one, you won't know which alteration is the consequence. Small things can add up: It's not just headlines that matter, but button colors and photo sizes as well.

Jumping the funnel occurs when you change or skip parts of the funnel while testing. A person's life experience or past project results may tempt them to jump the funnel. Have a wonderful idea that you want to try right away? Don't.

Wait. Let the first test run. If your amazing idea is in the middle of the funnel, wait until you've tested the pieces higher up. Jumping the funnel usually leads to redoing work. Decisions made without data are difficult to justify later.

If your funnel is complex, test individual components separately from the main funnel process. Product interest, pricing, headlines, and landing sites can all be explored and tried separately. It is advisable to test different funnel portions during your marketing campaign. Some marketers skip steps or jump the funnel based on strong results from smaller testing. This causes future issues. Caution. Intense discipline.

If the test does not go smoothly, stop it and restart. A compelling advertisement distributed at certain times of the day leads to a landing page with a relevant hard-to-refuse offer, followed by a tested shopping cart with a great user experience, allowing you to relax.

Working funnels don't indicate the testing is done. Now comes the tedious task of managing, testing, and refining the funnel. Jumping about the funnel now feels a little safer. Maybe you love your online shopping cart, but the landing page may use some work. You may now make changes because you know the funnel works. One part at a time.

We get lazy, and our experience tells us we can skip steps. This will steadily ruin your system, and you will wonder how it got so bad. Recognize it?

Rather than removing steps, it is better to integrate them, establish new procedures, and employ better tools. Examine your workflow, sales, and marketing funnels again.

If your sales funnel isn't working regardless of the medium, creativity, call to action, USP, or price, it's time to reevaluate your product or service. Jumping the funnel may never reveal what works and what doesn't.

Targeting a Generic, Broad Audience

If your target audience is quite broad, it suggests that your targeting criteria are not particularly rigorous and

that the demographic you've identified has many more listeners than your budget can successfully reach. While this is fine for reach campaigns in which you want to engage as many users as possible within your campaign budget, it also means that users will most likely only see your ad once. In contrast, with an awareness campaign, you would most likely want the same listeners to hear your ad many times in order to reinforce the message. You may control this by restricting your audience. This can be done by using stricter targeting parameters such as geographical targeting, demographics by age or gender, or calculating what content your target audience is most likely to engage with.

Using Incorrect Ad Placements

You now understand your Facebook ad placement choices and can select the finest places for your advertising. To select your ad location, follow these steps:

- Navigate to Ads Manager.
- Select your ad goal (not all goals are compatible with all ad placements).

- Navigate to the placements section (automatic placement is preselected).
- Choose manual placement.
- Uncheck the placements where your ad should not show.
- Choose the platforms on which you want your adverts to show (for example, desktop and mobile devices).
- Deselect (if appropriate) the platforms where you do not want your advertising to appear (Facebook, Instagram, Messenger, Audience Network).
- Save your adjustments.

Measuring the Wrong Metrics

Although I'm positive about allowing as many metrics as possible, having too many might be daunting if you don't grasp what each number means. The inclination is to go for higher numbers if each metric is not understood. Larger statistics are frequently beneficial, but they might lead you astray from your company's aim if they point in the incorrect direction.

If you get caught up in the numbers game, you may resort to strategies that increase your traffic, following, and everything else. However, more isn't necessarily better. If you want to grow sales leads, your approaches will be different than if you want to increase followers. While increasing your followers may result in more sales leads, you may be wasting a lot of effort if you're monitoring the incorrect KPIs.

When the incorrect measurements are measured, the wrong techniques are implemented, which leads to the wrong results. If you place a high value on metrics like followers and visits, yet they don't tell you anything about whether or not you're meeting your business objectives, you may be wondering why your social media strategy isn't working. Begin with your company objective and then consider how social media may help you achieve it. Then search for the correct Metrics to inform you if you're on the correct path.

Targeting the Right People With the Wrong Message

We've seen a variety of content types utilized to increase engagement and results on Facebook pages, including educational posts, motivational posts, give-aways, and sales posts. When compared to those that inspire a genuine dialogue, these ones today fall flat. When developing your content strategy for your website, maintain the notion of having a dialogue at the forefront of your mind.

Setting the culture of your page is the first step in promoting meaningful dialogue. Most people believe you achieve this simply by stating the topic of your page, however, this is not the case. Permissions are required. Identify something that your target audience feels bad about and give them permission to feel guilty about it.

If your target audience member is a mother who has launched a business, for example, many moms feel bad about how much they have to work, even if they work from home. Your page may offer them permission to be human and accept them for not being able to control all elements of their life flawlessly, giving them a sense of belonging within your community.

People want to feel like they belong, that they are welcomed and that any flaws they find in themselves are forgiven.

Once you've defined this permission for your website, it's critical to express it effectively through your content. Video, for example, is an excellent method to convey this permission since people can see your face and hear you speak. Upload a video and pin it to your highlighted area so that it is the first thing visitors see when they visit your page.

Targeting the Wrong People

Your first aim should be to target low-hanging apples, or those you know are interested in you. I know it's a lame argument, but it's worth remembering when you're about to spend a small sum on Facebook ads. Who might these apples be? Consider the following:

- your current clients
- persons that responded to a lead magnet recently
- your mailing list (which you've refined and kept warm with a steady stream of interesting mailings)
- people who have viewed your videos or clicked on your Facebook ads
- people who have been browsing your website or spending time on a specific page

Targeting the Right People With the Wrong Types of Facebook Ads

People are responding to your amazing video material on your website or Facebook page. You decide to retarget them with an ad using static pictures or a click to website ad.

Why aren't you putting up a video ad to this target demographic if you know they've responded to video before? It's not exactly rocket science (Social Media Examiner, 2018).

Your Ad Set Budget Is Too Low

Businesses are doing everything they can to save costs now more than ever. That also implies that many of them are reducing their advertising budgets.

To be honest, you can do a lot with Facebook advertisements on a limited budget. However, going too low may result in you not reaching the desired range.

The budget for your ad is created at the advert set level. The budget will be heavily influenced by the kind of bidding (automatic or manual) and the time frame you select, for example, daily/lifetime (BlueWaterMarketing, 2021).

Relevance Score

You're already really clever. You may already know your AdWords Quality Score. You know it has the capacity to impact where your ad appears, how many times it appears, and how much you pay. And you're aware (or should be) that a single point difference in either way can result in a 16% cost differential. It turns out that Facebook offers a comparable statistic in the Relevance Score of an ad. AdEspresso did a fast test of over 100,000 advertisements when this was originally released to evaluate how an ad's Relevance Score is associated with the effective Cost Per Click (CPC) and Click Through Rate (CTR). They launched two campaigns with identical ad design against each other to test and debunk. The first was to a random audience, whereas the second was to a specified bespoke target.

The one that was not well targeted had a Relevance Score of 2.9 and a cost per click of $0.142. The identical ad with superior targeting received an 8 Relevance Score, resulting in a cost per click of only $0.03, winning them four times as many clicks.

As a result, the Relevance Score differs from the AdWords score in that it does not 'rate' your design or text. Instead, it considers audience targeting to assess how relevant your message is to the demographics of individuals you're attempting to reach.

On the other hand, a good or average design and text with excellent audience targeting might nevertheless obtain a high Relevance Score (and thus, lower cost per click).

This score may be found by heading to one of your ad campaigns, scrolling down to a specific Ad Set, and then searching in the lower right-hand corner. The score is one out of ten, with one being awful and ten being excellent.

Awesome. But how do you go about fixing it? Should it be changed? Should it be improved? This is how: get a message that matches (Wordstream, 2021).

Matching Messages

The most effective internet advertising strategies combine a few essential elements or factors to offer visitors a unified, consistent experience from start to finish.

Message match (as it's known in the industry) implies that the key term you're targeting should be in the headline of your ad, as well as the headline of your landing page. This applies to Facebook ads and also Google Ads.

In actuality, it's a little more advanced, but you get the idea. To take an Unbounce example, they have an ad with great usage of clip-art, and you opt to click on it and go to the landing page.

The headlines, taglines, and graphics are all consistent. That is called message matching. It's so simple? Sure. However, many people continue to get this wrong.

One informal research conducted by Oli at Unbounce, for example, discovered that 98% of 300 distinct landing pages did not accurately align message matches.

Improving message matches in AdWords can assist raise your Quality Score, allowing you to pay less per click or lead.

Being attentive to message matches on Facebook may help you raise your Relevance Score, allowing you to pay less per click or lead.

In this situation, though, audience targeting is more important than keywords. That implies advertising should (and must) be generated (and then targeted) to as many individuals as feasible. (Such as distinct buyer personas.)

The wording in your content and the graphics you utilize should be linked with the pain areas or preferences of that specific persona.

Then you may utilize finer targeting options, such as interest overlapping, to accurately target advertising to those who are interested in both Interest A and Interest B.

Generic, one-size-fits-all advertisements and audience targeting reduce your Relevance Score and boost your expenses. And poor message match might affect conversions back on your site as well.

Ads Don't Align With Your Landing Page

You may have created the ideal ad but if your landing page is flawed, you will lose the transaction. Landing pages must be visually consistent with the ad, which means that the graphics, language, appearance, feel, and tone must be the same or comparable.

A well-designed landing page should incorporate social proof, address any concerns, utilize clear and short headlines that are easy to skim, and feature pleasant, error-free body text.

- Add trust indicators and make the call to action obvious and repeated across the page. In fact, everything on your website should lead to that action; that is the primary purpose of your page.
- It is also critical to reduce the number of connections. You want consumers to click on the CTA button rather than navigate away from your website (or worse still, to another website due to Google ads lurking on the page).
- On your landing page, generate a heatmap. Examine where people click and where they don't click. People will sometimes click on a component that does not have a link. This may be a simple repair.
- Create a scroll map to see where users leave the landing page. Shorten your page to that length and place your strongest CTA where users drop off. You should see a significant increase in conversions straight away.

- Examine user recordings. Nothing beats observing actual user behavior. What entices individuals to enter your funnel? Where are they attempting to go? Leverage the user's natural tendencies as much as possible.
- Run a series of A/B tests to identify and resolve any issues. Also, for your landing page, try 5-7 various titles. This can easily result in a 30% increase in conversions.

Do not expect immediate profits from ads, manage your expectations and realign your strategy. You will be well away on your road to ad-success soon.

8

SCALING

By now, the Facebook Ads Manager is no stranger to you any more! By now you've put up your first ads. You've learned a little about how to get new customers. It's time to start making real money with Facebook ads!

Scaling your Facebook ads will take your advertising to the next level. It allows you to boost your best campaigns and make more sales. You may investigate the following points:

- Add money to your ad spend.
- Make sure that your campaign budget doesn't get too big.
- Try new audiences that look like your audience and try new interests.

- Target more locations.
- Use your best ads again.
- Target a very wide range of people.
- Avoid including too many people at the same time.
- In the same campaign, you can change the offers.
- Keep the learning phase going at full speed.

When you scale ads, you get the most out of your best-performing ad campaigns. You do that by making some changes. Make sure you spend more money on ads, look for more high-performing audiences, or work on your creative. The goal is to get more leads or sales from the campaigns that are already working well for you, so you can keep them going.

SCALING THE METRICS OF THE ADS

Test New Audiences

▷ **Look-alike Audiences**

Facebook look-alike audiences are totally new audiences Facebook may generate by locating like-minded people from a given seed audience. For example, you may have Facebook generating a look-alike audience from a CSV list of consumers you submit. Look-alike audiences are immensely effective for discovering new viewers.

▷ Interest Audiences

Targeting new interests might help you reach new audiences and expand your Facebook ad campaigns. This technique is about discovering new interest audiences to target.

Build new ad sets with your top-performing advertisements once you've identified new audiences. You may need to design new ads to properly target your new viewers.

Increase Budget

The daily budget is increased by 20% every three days. Exceeding the threshold will reset learning and negatively damage outcomes.

Then you apply a large budget allocation depending on your intended cost each purchase and how much you need to spend to get out of learning in seven days. Remember that your budget determines your audience reach, the time it takes to see results, and your learning curve. Slow scaling requires a high daily budget, whereas quick scaling requires duplicating the existing campaign and leaving it activated until its efficacy starts to decrease.

Target More Geographic Locations

Targeting a broader location is a direct way to size your Facebook campaigns. If you don't need to specifically target only people living in a very specific area (for example if you own a small deli in a specific area), you can expand your target audience geographically nationally and even globally. After running many ad campaigns in one nation and gathering enough data to know what kind of advertising converts the best, you may use your insights to broaden your region targeting. If you've had success targeting in the U.S., for example, you might try running advertisements in other English-speaking nations where you see promise.

Here are some pointers for seeking innovative locations:

- Make some calculations to ensure that shipping to new countries or states is worthwhile for you.
- Be mindful of any cultural variations and adjust your adverts correspondingly.
- Begin with a lower ad budget and gradually expand your ad expenditure.

When Do You Need to Scale Down Your Facebook Ads?

As soon as possible! The moment your Facebook ad campaigns start making money for you, you can start thinking about how to make more money with them. Some ads that work are worth scaling, but not all of them are. Look for ads that have a good return on your money spent on them. In our first tip, we explain how to figure out Return on Ad Spend (ROAS).

Let a new Facebook ad run for at least three or four days before you decide if it's good or bad. This is the time Facebook needs to make sure your ads are delivered in the best way possible. Make the changes to the ad during this time. If you do that, you will change the Facebook algorithm.

When I Run Facebook Ads, How Can I Get More People to See Them?

Facebook ads can be scaled in two ways: horizontal and vertical scaling. Vertical scaling means changes to the budget, while horizontal scaling involves a lot of different ideas.

▷ Facebook Ads That Are Vertically Scaled

In order to increase the size of your Facebook ads, you just need to spend more money on them. Add to your ad budget a little at a time. This way, you'll also give

Facebook's algorithm time to adjust to your bigger ad spend and make sure your ads show up to the right people.

▷ Facebook Ads That Are Scaled Horizontally

If you want to make your Facebook ads bigger, you can also make them bigger horizontally. You can use it to get even more from Facebook campaigns that work well. With this method, you can make changes to multiple ad sets, change your audience targeting, and make changes to your creatives.

When you start horizontal scaling, you have to learn a lot of new things first. If you want to build a strong Facebook ad account structure and get great long-term results though, this process is the best way to do it.

Your Ad Budget Is a Simple Way to Grow Your Ads

- Use ROAS to figure out if it's worth it to spend more money on ads. You can figure out your ROAS for a whole campaign, an ad set, or just a few ads. The best results will come from scaling Facebook ads that are shown between 1.8 and 4 times a day.
- Use data from the Campaign Budget Optimization (CBO) tool to get a better idea of

how much money you should spend on your campaign.

- A Facebook ad campaign can have a set budget for all of its ads. You can also set individual ad set budgets to keep track of how much money you spend on each one.

- The second method works well if you want to be more in charge of delivering specific ad sets in your campaign.

- Among other things, you could have ad sets with different optimization goals or bid strategies in them. Ad sets with very different audiences could also be possible, as long as they aren't all the same in size. It's good to be in charge in these situations.

- There is, however, a lot of time and effort that goes into managing your budget this way.

- There are likely many things on your mind already as the owner of an online business. Adding one more thing to deal with isn't something you need.

- Campaign budget optimization helps to relieve some of the stress that comes with running a campaign There isn't much you need to do to set a budget for your ad campaign. Facebook will divide your money up between your best-performing ad sets.

▷ **The Advantages of CBO:**

- Simplified campaign setup and management: your money is automatically distributed.
- Efficiency: You get more out of your ads for less money when you use this method.

▷ **Things to Keep in Mind About CBO:**

- A campaign with two or more ad sets is the best one for this to work with.
- You can set a budget daily for what you can spend.

It's likely that you won't spend the same amount of money on each ad set. You could even spend 90% of your ad money on one ad set if it was the best one. Remember to look at your results at the campaign level, not the ad set level, because this is how you should look at your results.

Use CBO to Make It Easier to Grow Your Facebook Ads

There are two methods to establish your budgets for Facebook Ads. The first is at the ad set level, where you may specify a budget for each ad set inside a specific campaign.

CBO, on the other hand, is a (relatively) new Facebook advertising tool that lets you specify a daily budget for a whole campaign rather than individual ad sets. To go deeper, CBO uses Facebook's technology to allocate your spending among ad sets where conversions are most likely to occur at the lowest cost. Rather than keeping a consistent expenditure per ad set, Facebook will pool the cash into the ad sets that are most likely to produce results on any given day. What's the big deal about CBO? Why do Ad Managers need to learn this strategy as soon as possible?

While this may appear to be a simple adjustment, using Facebook CBO demands new approaches as well as new plans to be lucrative.

So, where do you begin? Let's start with the basics of how to put up a CBO campaign before going into sophisticated methods.

The first step in creating a new campaign is deciding on an aim. We've picked the conversion target in the example below. After you've chosen your aim, you'll be able to pick CBO before proceeding to ad set construction.

When you turn on Facebook CBO (to optimize your budget across ad sets), you'll be able to enter your daily budget and choose your bidding method.

Facebook will automatically choose the "lowest cost" bidding approach, but you will have three more options:

- Cost limit
- Bid limit
- Target cost

When we initially test an audience, we usually go with the cheapest option. We try the other three if the results are poor or if the client has very aggressive CPA targets that the cheapest option cannot reach. Of course, if an ad package fails to meet our expectations, we will disable it. At the ad set level, there are also a few complex choices to pick from. They consist of ad schedule and distribution mode.

Which Should You Use: Facebook CBO or Ad Set Budgets?

You may still pick between campaign budgets and ad set budgets. Most crucially, when should you use Campaign Budget Optimization over ad set budgeting?

While it is critical to test, test, test, we have discovered two major rules of thumb for when each of the individual strategies works well.

Use Ad Set Budgeting for Testing New Audiences

Ad set budgeting is, in most cases, the most effective strategy when testing new audiences. It guarantees your budget will deliver impressions to people within a specific set timeframe.

Use CBO for Ready Tested Ad Sets to Scale

Once you've done dozens of new tests and found your top-performing ad sets, including them into a CBO campaign is both a great practice and an incredibly effective scaling approach. We've tried a few different approaches to scaling campaigns using CBO. There are two ways that get the best results:

- Best Ad Sets
- Look-alike Audience + High Interest + Large Audience

Both of these structures are described in detail below, along with two more that you may (and should) try out.

Models of Facebook CBO Campaigns That Worked for Us

We've hammered out best practices for four CBO campaign structures that all Facebook marketers should put to the test:

- Guiding Ad Set (Facebook recommended)
- Best Ad Sets
- Look-alike Audience Ad Sets at 1 - 5%
- Look-alike Audience + High Interest + Large Audience

The structure you should adopt is determined by your goals and budget. And, as always, if one doesn't work, try another.

▷ Facebook Recommended Guiding Ad Set Structure

If you're new to Facebook CBO, this is a wonderful place to start. When utilizing this structure, you should duplicate your best-performing ad set into a new CBO campaign. Next, inside that same campaign, build a new broad audience ad set targeting your customer avatar with very minimal specifications such as age range, gender, and geographical area. That's all. There is no interest targeting or audience stacking. This CBO campaign contains two ad sets.

If you obtain at least 50 conversions in one week, your Facebook Pixel will really teach your broad audience "ad set" to discover your consumers. This enables the algorithm to undertake the grunt work.

According to Fetch and Funnel, this structure works best when you have a significant enough budget to obtain at least 50 conversions in one week.

▷ Best Ad Sets

So, what do you do if your budget does not allow you to train the pixel? Simply combine all of your top-performing ad sets into a single CBO campaign.

Begin with 5-10 of your most successful sets. You can use fewer, if necessary. What matters most is that they work properly for you.

This framework has shown to be quite effective for us, particularly when scaling campaigns. What we appreciate about it is that you can plainly see which ad sets are still driving results and simply switch off ad sets that are no longer performing.

This structure provides marketers greater control over the audiences Facebook targets.

▷ Look-alike Audience Ad Sets at 1-5%

With this design, it's generally preferable to start with the picture. Fetch and Funnel tested this layout with up to ten ad sets, but had the best success with only five. More testing is at your own risk.

When putting this framework together, make careful to keep each audience separate from the others to avoid audience overlap. As a result, the 1% look-alike audience will be removed from the 1-2%, 2-3%, 3-4%, and 4-5% look-alikes... and so on for each ad set.

Though the setup is a little complicated, it may yield excellent returns if it works for your company. We recommend giving it a go, especially if you've had difficulties expanding your prospecting beyond a single look-alike demographic.

▷ Look-alike Audience + High Interest + Large Audience

Give this bad boy a go if you're getting the appropriate amount of conversions through your pixel each week. It is the ultimate scalable campaign and is extremely successful for lead generation and conversion initiatives.

To establish this structure, divide all of your best-tested audiences into two ad sets—one for interests and one for look-alikes—that are sometimes referred to as "Super Groups." Then, to enable the algorithm to do its thing, you construct a third ad set composed of a large demographic category.

Because each of the audiences has previously been tested, this campaign has a high success rate. It also employs look-alike and interest targeting to direct the broad audience ad set. We prefer to start with a testing campaign, following the framework provided in the next section.

In the same ad, we test both look-alikes and interests. We add that look-alike or interest to one of the two super ad sets after testing the ad set and seeing results that satisfy your intended goals. This keeps the audience engaged and growing in a safe and regulated manner.

9

SUCCESS STORIES

You are probably thinking to yourself that you are too old, too new with this, or too unlucky to be successful with Facebook ads. *I can barely power a computer, only lucky people will make any money etc...* These success stories will surely change your mind and boost your confidence to start. Company names are used for illustrative purposes only and have been changed to maintain anonymity.

INSPIRING FACEBOOK ADVERTISING SUCCESS STORIES

ABC Couture Double Revenue

ABC Couture needed a strengthened international brand identity and recognition. To achieve this, their

CEO and designer-in-chief turned to Facebook advertising.

ABC created a Facebook marketing campaign aimed at a specific market. They used Facebook's advanced targeting techniques to target rich women aged 18 to 60 who were interested in distinctive and attractive apparel. Along with Facebook's targeting capabilities, they also utilized Look-alike Audience to reach new potential consumers who shared characteristics with individuals who had previously reacted to their advertising. They increased the reach of their posts by boosting them and delivering advertising to their Custom Audience.

ABC Couture, however, needs more benefits from Facebook for business than a highly focused audience. Throughout their campaign, they painstakingly did A/B tests, comparing their adverts' photos against photos, and videos to videos, to determine which ads with which aspects garnered the most likes. Then, they modified their advertising plan depending on the results of their testing, vastly enhancing their campaign each time—and their efforts were rewarded.

ABC's video advertisements had over 20,000 views, and their campaign directed thousands of women to their website. Their sales climbed by 55% in six months, more than tripling their revenue.

Facebook has assisted ABC in becoming a worldwide fashion brand that caters to an international market. ABC Couture's effective Facebook ad campaigns place their company in the spotlight.

Eezi Soda Company Reached Millions Across Devices

Eezi Soda Company utilized Facebook advertising to increase brand recognition in South Africa. Eezi Soda Company, Facebook, and its agency Hello Computer collaborated to design a multimodal campaign that would eventually reach the number of people Eezi Soda Company sought.

Eezi executed a 3-month campaign that targeted millions of consumers across various sorts of mobile devices using Facebook capabilities, such as reach and frequency purchasing, Carousel, Canvas, and Video advertisements. The intriguing storyline in the company's video and canvas commercials drew in viewers. Eezi employed Facebook slideshow advertisements, which can accommodate devices of any internet connection, to guarantee that the firm reached even those who owned devices with inferior bandwidth.

The efforts of the beverage firm were not in vain. Eezi expanded its individual reach by 42% between March 1 and May 30 in that year, despite spending just 3.5% of its budget. It also enhanced the brand's effect and total campaign reach by 96%. Eezi discovered that Facebook advertising was three times more cost-effective than TV advertising and six times more cost-effective than other digital video advertising throughout the campaign.

Eezi was thrilled with the outcomes of the initiative in collaboration with Facebook. In their media statement, they explained that their collaboration with Facebook on this initiative was highly fruitful. Not only did they effectively reach their core audience and achieve the major objectives of landing the new brand relaunch, but they also acquired priceless information about Facebook's reach within the media mix through the first cross-media research in Africa, which would affect how they proceed in the future.

What Will $5/day on Facebook Advertising Get You?

I'll get straight to the point: If you're presently spending money on online advertising and not on Facebook, you should seriously consider switching your bucks over to Facebook. With the enormous surge in enthusiasm in advertising and promoting products and services on Facebook, Benschop conducted a modest experiment to explore how a $5 investment in Facebook Ads may produce results.

The $5 ad expenditure was evaluated for various goals such as number of page likes, website clicks, and impacts on boosted content. Here follows an example that compares to the outcome that they got for every $5 they spent on each of the activities listed below (Benschop, 2017):

- Likes on the page - 10 likes per day
- Page Views - 1 per day
- Post was promoted, and 790 additional individuals were reached as a result.

So, even on a minimal budget, there is undoubtedly some form of return for enterprises.

Here's what you're aiming for:

- As you can see, they were focusing on folks who are interested in social media, rather than the company's current fans (they wanted only cold traffic).
- Furthermore, they paid great attention to the ad design for the Page Likes campaign to create a killer Facebook Ad.
- Their ad aimed to be highly socially orientated (since it shows the team), which is why it performed well for a 'like' campaign.

So, to get the most bang for your buck with Facebook Ads, even on a tight budget, consider focusing on your target group and writing ad language that reflects or resonates with them.

▷ **Lessons Learned:**

- If you want to test the market of Facebook Ads or develop discussion for your product, a little expenditure may also help. Not only can it raise awareness for your company, but it may also result in useful traffic and revenue.
- To be successful with Facebook Ads, you don't need a significant budget. Regardless of your goal, you may still obtain high traction if you target and manage your advertising properly.
- Always exclude your existing page followers when aiming for page likes.
- Ensure that the ad picture/creative you choose corresponds to the ad's purpose.

Facebook Ad Success With Just $1 per Day

Similar to the above experiment, a case study by Brian Carter, a notable Facebook Marketing and Advertising Expert and Bestselling author of the book *The Like Economy*, showed that paying as little as $1 per day in Facebook Ads may provide big results.

He was able to reach 120,000 individuals, or 4,000 people every day, by continuously investing $1 per day for 30 days. He is an avid user of most advertising networks, and this is the cost he discovered for reaching 1,000 individuals through common advertising channels.

Facebook Ads are far less expensive than traditional advertising alternatives (newspaper, television, etc.), however it has fallen behind its online competitors (Adwords and LinkedIn).

The goal of this case study or experiment was to demonstrate that even with a very limited budget, Facebook Ads may be helpful.

▷ **Lessons Learned:**

Budget is or should not be a roadblock for virtually any business. Isn't it true that most companies can afford to pay $1 each day on Facebook?

Even if you are effectively spending on other channels for traffic or lead generation, it doesn't harm to spend a tiny amount of your budget on Facebook Ads. You may receive the same amount of traffic, but the overall cost will be far lower than the other options.

How an Advertising "Disaster" Can Actually Turn Into a "Win"

This case study is intriguing since it began with a failure before they had their winning aha-moment.

Nisha Mitchell launched a Facebook Ad campaign for a client who intended to market a high-priced $980 home improvement program to women aged 30-40 across the United States.

The plan was to use Facebook Advertising to create an email list and then show them ads for a free webinar that would lead to the package she wanted to sell.

When they ran the advertisements, they performed well, but the majority of those who engaged with them were not the intended audience.

She expected women in their 30s and 40s to be more interested in her program, but women over 50 were the most engaged with her ad and material.

The first several days only produced two sign-ups at a cost of roughly $28 a piece. As a consequence, she adjusted the ad targeting and wording to match the prior results.

The second run drastically reduced the cost/lead, lowering it to $4.43 per lead.

After numerous rounds of testing and adjusting with targeting and ad copy, she was able to collect 400 leads for $508 and several delighted clients who went on to purchase the program.

One of their best-performing advertisements was shared 15 times, indicating that the ad content was speaking to the customer and was effective.

It is critical that you continue to evaluate and optimize your campaign, especially at the beginning. Angela was able to cut the cost/lead from $28 to $1.28 in the end by continuously optimizing the campaign based on the results she was receiving.

Unless you have a high budget, test your campaign in modest budget/test runs until you uncover what works, otherwise you risk spending a lot of money with minimal returns.

▷ **Lessons Learned:**

There is no restriction on the type or cost of the goods you are selling; You may even sell a high-priced product using Facebook Ads.

Run some basic ad tests to fine-tune the audience before investing more money where you can obtain the most gold.

Your perception of the optimal audience may not always be right.

Sometimes the simplest things are the most effective. The CTR and leads were doubled by replacing the word 'Webinar' in the ad copy to 'Workshop.'

How Verando Got 500% ROI From a $20 Facebook Ad Budget

This case study is very valuable for individuals and enterprises that do not have their own products but promote affiliate items. We frequently see our clients use Facebook Ads to promote affiliate deals, but they appear to struggle with developing an effective plan.

One of the most common errors they make is failing to use the landing page properly.

Sending traffic from Facebook is not difficult or impossible. You will still be able to receive the necessary traffic if your targeting is correct.

However, if you're marketing affiliate items on Facebook, direct them to a 'bridge' page first, rather than a sales page. This is what Verando's Social Media team did. They were pushing the affiliate offer with two landing sites.

- A single page with an explanation of the offer and a Free Trial button that directed them to the sales page.
- One with an opt-in form (just for email) that leads to the sales page after submission.

They paid a total of $20 on advertisements on both pages and earned $100 in commission and 60 quality leads.

Despite the fact that the sales came from the first page, he was able to get meaningful leads from the second ad and landing page.

They did not only make $80 more than they spent, but they also expanded their list, which they may use in the future. Almost 120 of the 200 persons that clicked on his Facebook Ads clicked on the affiliate link. As a result, bringing good quality traffic undoubtedly works in your advantage.

▷ **Lesson Learned:**

- To get the ultimate results from your campaigns, don't simply focus on the Facebook Ads, prepare ahead of time where you'll send the traffic and how you'll capitalize on it.
- Make several funnels and objectives using various ad sets and landing sites. In this

scenario, he tested two advertisements on two separate landing pages, providing him with numerous segmented audiences.

So there you have it, an in-depth examination of the top chosen Facebook Ads Case studies from practically all sorts of businesses with varying goals.

Reviewing the case studies above, it is clear that the effectiveness of Facebook Ads is dependent on a variety of factors. However, it is apparent that budget is not the most important issue, and you may achieve it even with a small budget.

If you have learned something new or enjoyed the content of this book, please write a review on Amazon. https://www.amazon.com/review/review-your-purchases/

CONCLUSION

Everything I've been great at in life—everything you've grown really strong at in life—has a lot in common with mountain climbing. It's not such a huge thing if you've mastered a talent, whether it's climbing a 10,000-foot peak or earning a million dollars a year. You may not even develop a sweat on a good day. Most folks are put off by the first 50 feet—making that first dollar of genuine profit. Once you've passed that stage, the competition begins to diminish drastically. Soon, you'll discover that you're not just familiar with it, but also enjoying it. That is how a serial entrepreneur is born: hooked to the experience and unable to quit.

Friends, I've learned to know that horror, that chilly sense of vulnerability and panic, is a sign that what I've just begun is significant. That is what should be done. The fear indicates that the journey is worthwhile.

Yes, you may experience dread if you've embarked on something completely idiotic. But you'll experience that anxiety whether you start a new business, try to mend a broken relationship, or make a vital phone call. Nothing significant transpires unless you do it.

The winner gets it all, and you are the winner when you choose to fight fear and turn it into an asset. Now that you have all of the tools, put them to use.

If you have learned something new or enjoyed the content of this book, please write a review on Amazon. https://www.amazon.com/review/review-your-purchases/

FACEBOOK ADVERTISING GLOSSARY

Ad

A Facebook Ads Manager advertisement that you build to promote your e-store, Facebook page, services, goods, and so on. It often consists of a picture, content, and a call-to-action button.

Ad Auction

Unfortunately, when you make an ad, it does not immediately appear in the feed of a Facebook user. Simply put, it is auctioned off against other similar commercials. Facebook uses ad quality, bid, and expected reaction rates to determine which ads in the bid to show to Facebook users and when.

Ad ID

It's a special number assigned to each of your adverts.

Ad Set

It's a special number assigned to each of your adverts.

Ads Manager

Facebook Ads Manager is a service that allows you to create, monitor, and execute Facebook ads, and also manage all of your payment and billing details for Facebook ads. The Ads Manager includes an analysis dashboard where you can track the effectiveness of your ads.

Adds to Cart

Whenever anyone (who came to your website via your Facebook ad) submits an item to the cart, it appears in the reporting section as an Add to Cart event. Add the Facebook Pixel to your site to see this in your ad reports.

Amount Spent

It's a feature in a reporting table that shows how much revenue you spent on an advertisement, campaign, or ad set.

Audience

An Audience is a specific group of people you decide to target using your Facebook ad.

Audience Network

A group of Facebook's partners. There are numerous mobile app and web publishers who will display your adverts on their applications and websites.

Average CPC

The average cost-per-click is a figure that displays how much each click on your ad costs you. CPC is computed by dividing the entire ad spend by the number of clicks obtained by your ad. When calculating ad costs, use this statistic as a guideline.

Bid Strategies

A bid strategy is a method used by Facebook to bid on your behalf in an ad auction. When manually setting the bid, select one of the following strategies: lowest possible cost, bid cap, cost cap, greatest value, target cost, or min Return on Ad Spend.

Broad Categories

Facebook provides a list of broad targeting categories to make ad targeting easier for you. You can group your users by interests, Facebook likes, apps installed, and so on.

Budget

The budget defines the maximum amount of money you want to spend on each campaign or ad set. If you have enabled Campaign Budget Optimisation (CBO), you then will have to manage your budget at the campaign level, not at the ad set level. You set the amount of money to be spent either over the lifetime of your ad (or campaign) or on average for each day.

Boosted Post

You don't have to run ads to create brand awareness or gain more followers. You can also try boosting (sponsoring) a post on your Facebook page. By boosting a post on your Facebook page, you can make it visible to Facebook users, who have not liked your page. However, if you want more optimization options, create ads on Ad Manager.

Campaign

All ads you create on Facebook go under separate ad sets. All ad sets fall under different campaigns. Here's an example of the structure a clothing store might have in Ads Manager: Campaign (Black Friday) -> Ad set (Women) -> Ad (Black Friday Women Promo 50% Off).

Campaign ID

A unique number given to all your campaigns. The number is visible in the reporting section of the Ads Manager.

Campaign Spending Limit

You can set a maximum spending limit for each campaign. After the limit is reached, your campaign stops running.

Campaign Reach

Campaign reach refers to the total number of people who saw your ads in the campaign.

CBO (Campaign Budget Optimization)

Campaign budget optimization is set by default, and it manages the budget at the campaign level. Based on your chosen bid strategy, Facebook automatically distributes the campaign budget across different ad sets, this way optimizing the campaign's performance.

Clicks

Clicks are actions performed by people who viewed your ads and clicked on them. Clicks can be: post reactions, comments, clicks on a business's profile or picture, post shares, clicks to expand, clicks connected to campaign objectives (page likes, messages, website visits, etc.), and so on.

Connections

When choosing an audience to target with your ad, you can choose to include or exclude people who like and follow your page. Those people are grouped as Connections.

Conversions

Actions completed by the customers, such as add to cart or purchases on your website, are conversions.

Cost Per Action

The average amount of money you pay for one conversion.

Cost Per Click (CPC)

The average amount of money you pay for a click on your ad.

Creative Hub

It's a tool you can use to create your ads on Facebook. Use it to also view ads created by other businesses and draw inspiration.

Click Through Rate (CTR)

CTR is a metric that measures the number of clicks your ad received per each impression. It's a kind of a success rate—the bigger the percentage, the better your ad is performing. Facebook calculates it by dividing the total amount of impressions by the clicks the ad got. If the percentage is low, tweak your audience, visual, or text to get better results.

Custom Audience

Custom Audience is a targeting option you can choose when creating a Facebook ad. To create a custom audience you need to import details about your customers, such as email addresses or actions made on your website.

Daily Budget

Daily budget is the amount of money you want to spend on an ad set or campaign per day. Each ad set can have its own daily budget.

Delivery

Delivery is the status of your campaigns, ad sets, and ads. The status is shown in the Campaigns page of Facebook Ads Manager and can be anything from active, to paused, or off.

Destination

The destination is the place you want people to land on after clicking on your ad or a CTA (call-to-action) button. It can be your business's website, messenger, Facebook page, or an app.

End Date

It's a date your ad set is scheduled to end. You set it when creating the campaign.

Frequency

A person may see your ad more than once on their Facebook feed. Frequency is the metric that defines the average number of times your ad was shown to the same person.

Impressions

The average number of times your ad was on your target audience's feed.

Interaction

If a user interacts with your ad in any way, it's marked as an interaction in the reporting section. The interaction can be anything from clicking your ad to watching a video. For example: 1 click and 1 comment by 1 person is marked as two interactions.

Lead Generation

It's an objective you can choose when creating your ads on Facebook. If you've set your campaign objective as lead generation, you can ask your audience to fill out a form with their contact details. The form doesn't have to be located on Facebook, it can also be on your website and tracked by Facebook if you use Facebook Pixel.

Lifetime Budget

It's an amount of money you've set to be spent over the lifetime of the campaign.

Link Clicks

It's a number of clicks on links in the ad that lead to a destination on or off Facebook.

Likes & Interests Targeting

It's a way to target Facebook users based on the pages they like and the interests they express on Facebook.

Look-alike Audience

Facebook creates look-alike audiences to help advertisers target audiences, similar to the ones that visit their website. To create a look-alike audience, you need to have a Custom Audience set up first.

Objective

The campaign objective indicates what you want to achieve with that campaign. It can be brand awareness or reach, app installs, video views, catalog sales, and so on.

On-Facebook Purchases

All purchases Facebook users carried out as a result of seeing your ad on Facebook-owned property (messenger, Facebook pages, etc.)

Page Engagement

A number of actions that Facebook users performed on your business page or a post, due to your Facebook advertising efforts.

Page Likes

The number of people who liked your Facebook page.

Payment Method

Payment methods are ways you can pay for the ads you create on Facebook.

Facebook Pixel

Facebook Pixel is a code snippet you can insert into your website's code. Pixel helps Facebook track all events related to your ads that take place off Facebook.

You can also use data from Facebook Pixel to optimize your ads.

Placement

When creating your ads in Facebook Ad Manager, you can choose where you want your ads to be shown on Facebook. You can choose to show them on News Feed, Suggested Video, Market Place, Right Side Column, and so on. You can also let Facebook decide where to show your ads with the automated placements option enabled.

Post Engagement

The number of actions (likes, comments, shares, etc.) people performed on your page post after they saw your boosted post or ad.

Potential Reach

It's an estimation of how many people are in your chosen target audience.

Promoted Post

A promoted post is a boosted post on your business's Facebook page.

Reach

It's the number of people who saw your ads once. Don't mix it up with impressions as impressions can include multiple exposures of your ad to the same group of people.

Relevance Score

Once your ad collects at least 500 impressions, Facebook scores it. The score is between 1 and 10, 1 noting low relevance, 10 noting high relevance. If your ad has a low relevance score, you might want to optimize it (tweak the audience, content, images, etc.)

Reports

The Reports section in the Ads Manager shows you the most important metrics of your ad and how it has been performing so far. You can choose which metrics to include and which not to.

Return on Ad Spend (ROAS)

In mobile and online marketing, this is an essential key performance indicator (KPI). It is the sum of money generated for every dollar invested on a campaign.

Social Clicks

Social clicks are ad clicks performed by someone whose Facebook acquaintance previously acted on the very same ad (e.g. Ann Adams likes this, Ann Adams follows this, Ann Adams is attending.)

Social Impressions

The amount of times your ad was displayed with social data (e.g. Ann Adams follows this.)

Start Date

The start date of your ad.

Status

The current status of your campaign, ad sets, and advertisements. It is either active (on) or inactive (off.)

Suggested Bid Range

When you create an ad on Facebook, it will propose a bid range for you to use. The bid is determined by the level of competitive pressure and the contents of your ad.

Targeted Audience

Based on the audience you target, this is the number of people your ad could reach.

Traffic

When developing your ads, you may select to target traffic. This goal aims at attracting as many individuals as possible to your website.

Transaction

This is a fee that Facebook charges you for your online ads. All charges for your adverts are listed in the Billing Summary tab of Facebook Business Manager.

Unique Clicks

This is the total number of persons who engaged with your ads. It should be noted that this only counts people, not activities.

Verification Hold

To confirm your payment method, a charge is made to your card. It is normally for $1.01 and will be returned to you within five days.

BIBLIOGRAPHY

AdEspresso. (n.d.). *Facebook ads examples - A curated selection of real ads to inspire you.* Retrieved May 8, 2022, from https://adespresso.com/ads-examples/?msclkid=2785aef6cec711ec90020b12da93ec7b

(2020, January 7). *The Facebook ads funnel guide – How to design a perfect Facebook funnel for your business.* StableWP. https://stablewp.com/the-facebook-ads-funnel-guide-how-to-design-a-perfect-face book-funnel-for-your-business/

Animalz. (2021, March 18). *11 expert Facebook ad design tips to increase conversions (+ examples).* AdEspresso. https://adespresso.com/blog/9-tips-perfect-facebook-ad-design/

Baltagalvis, A. (2016, March 30). *Don't touch that Facebook boost post button.* Agorapulse. https://www.agorapulse.com/blog/boost-post-facebook-advertising/

Barker, S. (2019, May 2). *9 step guide to creating a Facebook ads funnel.* Gist. https://getgist.com/creating-a-facebook-funnel

Barker, S. (n.d.). *9 steps to create a Facebook sales funnel that converts.* HubSpot. Retrieved April 24, 2022, from https://blog.hubspot.com/sales/facebook-sales-funnel

Blue Water Marketing. (2021, May 25). *Why are my Facebook ads not getting any reach in 2021?* https://bluewatermarketing.com/why-are-my-facebook-ads-not-getting-any-reach/

Benschop, C. (2017, February 17). [Case Study] *What Will $5/Day On Facebook Advertising Do For You?* LinkedIn. https://www.linkedin.com/pulse/case-study-what-5day-facebook-advertising-do-you-corey-benschop-

Claire, B. (2019, December 20). *27 Facebook demographics that matter to marketers in 2020.* Hootsuite Social Media Management. https://blog.hootsuite.com/facebook-demographics/

Coles, T. (2019, December 4). *30 social media content ideas and examples*

for brands. Hootsuite Social Media Management. https://blog.hoot
suite.com/content-idea-cheat-sheet/

Corrin, S. (2021, March 18). *How to create a Facebook landing page (6 super easy steps).* Seedprod. https://www.seedprod.com/how-to-create-a-facebook-landing-page/

Cyca, M., & Zarzycki, N. (2018, July 23). *The complete guide to social media video specs in 2018.* Hootsuite Social Media Management. https://blog.hootsuite.com/social-media-video-specs/

Denny, D. (n.d). *How to master Facebook ads.* Fetch & Funnel. Retrieved April 24, 2022, from https://www.fetchfunnel.com/how-to-master-facebook-ads/

D'Onfro, J. (2015, July 9). *Here's how much time people spend on Facebook per day.* BusinessInsider. https://www.businessinsider.in/tech/Heres-how-much-time-people-spend-on-Facebook-per-day/arti cleshow/47995030.cms

Full Funnel Marketing. (2016, December 12). *5 tips for a successful Facebook like campaign.* https://fullfunnelmarketing.com/blog/5-tips-successful-like-campaign/

Gotter, A. (2019, November 8). *What does a Facebook sales funnel look like?* Agorapulse. https://www.agorapulse.com/blog/facebook-sales-funnel/

Hoben, N. (2021, August 6). *Facebook video tips: 15 ideas for more engagement.* Search Engine Journal. https://www.searchenginejournal.com/facebook-video-tips/238911/#close

Hubbard, A. (2017, March 19). *Facebook ad not approved? How to prevent it from happening again.* Hubbard Digital Pty. Ltd. https://andrewhubbard.co/facebook-ad-not-approved/

Julija. (2021, January 22). *Scaling Facebook ads: 9 best tips to get sales.* Sixads. https://sixads.net/blog/scaling-facebook-ads/

Karlson, K. (2017, February 16). *How to create a Facebook like campaign.* AdEspresso. https://adespresso.com/blog/facebook-like-campaign/

Lawrance, C. (2021, September 17). *How to scale your Facebook ads: Two proven methods.* Social Media Examiner. https://www.socialmediaex

aminer.com/how-to-scale-your-facebook-ads-two-proven-methods/

LePage, E., & Newberry, C. (2021, October 19). *How to create a social media strategy in 9 steps (free template).* Hootsuite Social Media Management. https://blog.hootsuite.com/how-to-create-a-social-media-marketing-plan/

McLachlan, S. (2018, November 21). *How to create a buyer persona (free buyer/audience persona template).* Hootsuite Social Media Management. https://blog.hootsuite.com/buyer-persona/

McLachlan, S., & Newberry, C. (2020, March 8). *Facebook marketing in 2021: How to use Facebook for business.* Hootsuite Social Media Management. https://blog.hootsuite.com/facebook-marketing-tips/#TypesofFacebookposts

Mulvey, J. (2018, January 8). *How to create the perfect Facebook ad in minutes.* Hootsuite Social Media Management. https://blog.hootsuite.com/perfect-ad-facebook-minutes/

Neil Patel. (2017, March 18). *The step-by-step guide to creating a Facebook sales funnel.* https://neilpatel.com/blog/facebook-sales-funnel/

Newberry, C. (2019, October 2). *How to advertise on Facebook in 2020: The definitive Facebook ads guide.* Hootsuite Social Media Management. https://blog.hootsuite.com/how-to-advertise-on-facebook/

Newberry, C. (2021, July 15). *How to create a Facebook business page in 7 easy steps.* Hootsuite Social Media Management. https://blog.hootsuite.com/steps-to-create-a-facebook-business-page/

Page, M. (2018, November 7). *7 reasons your Facebook ads do not convert.* Social Media Examiner. https://www.socialmediaexaminer.com/7-reasons-facebook-ads-do-not-convert/

RevLocal. (n.d.). *3 myths about boosting Facebook posts.* Retrieved May 7, 2022, from https://www.revlocal.com/resources/library/blog/3-myths-about-boosting-facebook-posts

Shleyner, E. (2020, November 3). *19 social media metrics that really matter—and how to track them.* Hootsuite Social Media Management. https://blog.hootsuite.com/social-media-metrics/

Sides, G. (2017, October 31). *How to get more Facebook likes: 10 tactics that actually work*. Hootsuite Social Media Management. https://blog.hootsuite.com/how-to-get-more-likes-on-facebook/#strategy

Smith, B. (2021, December 2). *5 reasons your Facebook ads aren't working*. WordStream. https://www.wordstream.com/blog/ws/2016/10/25/facebook-ads-not-working

SocialPilot Team. (n.d.). *Is it worth boosting Facebook posts - When and how to do it*. SocialPilot. Retrieved April 24, 2022, from https://www.socialpilot.co/blog/worth-boosting-facebook-posts-when-and-how-to

The Revealbot Blog. (2020, May 18). *How to auto promote Facebook posts (yes it's worth it)*. https://revealbot.com/blog/promote-facebook-post/

Vega, N.D. (2021, August 2). *How to create Facebook page like ads to boost social proof*. Madgicx. https://madgicx.com/facebook-page-like-ads/

IMAGE REFERENCES

Alexas_Fotos. (2016, September 19). *Medios de comunicación social* [Image]. Pixabay. https://pixabay.com/photos/social-media-internet-security-1679230/

Altman, G. (2014, December 2). *Social media faces photo album* [Image]. Pixabay. https://pixabay.com/illustrations/social-media-faces-photo-album-550767/

Blomkvist, M. (2020, December 3). *Business plan schedule written on the notebook* [Image]. Pexels. https://www.pexels.com/photo/business-plan-schedule-written-on-the-notebook-6476808/

Buissinne, S. (2015, April 14). *Magazine image leisure* [Image]. Pixabay. https://pixabay.com/photos/magazines-reading-leisure-716801/

Dziuba, T. (2018, March 9). *Photo of laptop near plant* [Image]. Pexels. https://www.pexels.com/photo/photo-of-laptop-near-plant-927629/

Facebook Business. (n.d.). *Audiences* [Image]. Facebook Business. https://business.facebook.com/adsmanager/audiences

Geralt. (2017, January 18). *Medios de comunicación social* [Image]. Pixabay. https://pixabay.com/es/illustrations/medios-de-comuni caci%c3%b3n-social-1989152/

Geralt. (2022, April 18). *Empresario el marketing digital* [Image]. Pixabay. https://pixabay.com/es/illustrations/empresario-el-marketing-digi tal-7140598/

Henry, M (n.d.). *Tech group meeting flatlay.* [Image]. Burst. https://burst. shopify.com/photos/tech-group-meeting-flatlay?c=flatlay

Khors, N.D. (n.d.). *Contact us lettering* [Image]. Burst. https://burst. shopify.com/photos/contact-us-lettering

Mclean, E. (2020, March 21). *Person holding black Samsung android smartphone* [Image]. Unsplash. https://unsplash.com/photos/ Z41_IZ6Ctis

Mucira, J. (2020, May 21). *Social media* [Image]. Pixabay. https:// pixabay.com/illustrations/social-media-social-marketing-5187243/

Nepriakhina, D. (2019, March 31). *Vehicle beside wall with graffiti* [Image]. Unsplash. https://unsplash.com/photos/gGoi4QTXXBA

Pixabay. (2016, December, 22). *White Facebook scramble pieces* [Image]. Pexels. https://www.pexels.com/photo/advertising-alphabet-blog-close-up-267371/

Pixabay. (2017, August 27). *Facebook application icon.* [Image]. Pexels. https://www.pexels.com/photo/facebook-application-icon-147413/

Sanderson, K. (2019). *black-friday-social-media-post-black* [Image]. Pixabay. https://pixabay.com/vectors/black-friday-social-media-post-black-4606225/

Saura, J. F. F. (n.d.). *Street lights* [Image]. Pexels. https://www.pexels. com/photo/street-lights-802024/

Shopify Partners. (n.d.). *Free shipping wooden sign* [Image]. Burst. https:// burst.shopify.com/photos/free-shipping-wooden-sign?c=shipping

Solomin, D. (2021, November 5). *Facebook's transition to Meta — in 3D* [Image]. Unsplash. https://unsplash.com/photos/yIT9HO8UrPA

Tran, B. (2017, October 28). *Inspirational quote on a planner* [Image].

Pexels. https://www.pexels.com/photo/inspirational-quotes-on-a-planner-636243/